This unique publication summarizes fifty years of Russian research on shock compression of condensed matter using chemical and nuclear explosions. This information, and the equations of state derived from it, have important applications in physics, materials science and engineering.

An introductory chapter describes the importance of Russian experiments in a global context. The second chapter describes the experimental devices used. Following chapters summarize the results of experiments on pure metals, metal alloys and compounds, minerals, rocks, organic solids and liquids. The book covers experiments with pressures ranging from 2.5 GPa to 1 TPa using chemical explosives in laboratory conditions and to 10 TPa in underground nuclear tests. Attention is given to theoretical aspects, experimental problems and data analysis. The data in this book are quite unique as, with the cessation of large scale underground nuclear tests, it will be some time before similar pressures can be generated by alternative means.

This book will be of interest to condensed matter physicists, materials scientists, earth scientists and astrophysicists.

T0275561

Shock Compression of Condensed Materials

Shock Compression of Condensed Materials

R. F. Trunin
All-Russian Research Institute of Experimental Physics, Sarov

CAMBRIDGE
UNIVERSITY PRESS

CAMBRIDGE UNIVERSITY PRESS
Cambridge, New York, Melbourne, Madrid, Cape Town, Singapore, São Paulo

Cambridge University Press
The Edinburgh Building, Cambridge CB2 2RU, UK

Published in the United States of America by Cambridge University Press, New York

www.cambridge.org
Information on this title: www.cambridge.org/9780521582902

First published 1998
This digitally printed first paperback version 2005

A catalogue record for this publication is available from the British Library

Library of Congress Cataloguing in Publication data

Trunin, R. F. (Ryurik Fedorovich), 1933–
Shock compression of condensed materials / R. F. Trunin.
p. cm.
Includes bibliographical references.
ISBN 0 521 58290 3
1. Condensed matter – Effect of radiation on. 2. Shock mechanics.
3. Materials – Compression testing. 4. Materials at high pressure.
I. title.
QC173.458.R33T78 1997
530.4'16–dc21 97–3026 CIP

ISBN-13 978-0-521-58290-2 hardback
ISBN-10 0-521-58290-3 hardback

ISBN-13 978-0-521-01924-8 paperback
ISBN-10 0-521-01924-9 paperback

Contents

Preface   ix

1   **Introduction**   1

2   **Measuring kinematic parameters of shock waves**   16
2.1   Laboratory shock generators   16
2.2   Specific features of full-scale experiments   22

3   **Shock compression of metals**   25
3.1   Absolute laboratory measurements of kinematic parameters   25
3.2   Absolute measurements of metal compressibility in under-
        ground nuclear tests   33
3.2.1   Shock compression of iron in a pressure range of 4–10 TPa   34
3.2.2   Compressibility of copper, lead, and cadmium under a pres-
          sure of up to 5.1 TPa and molybdenum up to 1.5 TPa   43
3.2.3   Compressibility of aluminum in a range of 0.2–1.7 TPa   49
3.3   Comparison of laboratory data on metal compressibility to
        measurements performed in underground nuclear tests   52
3.4   Relative compressibilities of iron, copper, lead, and tita-
        nium   56
3.5   Compression of porous metals   63
3.6   Equations of state for metals   76
3.7   Compressibility of initially melted and supercooled metals   91

4   **Compression of metal compounds**   95
4.1   Shock compression of metal alloys   95
4.2   Compression of carbides and nitrides of metals   99
4.3   Metal hydrides   103

5   **Minerals**   106
5.1   Oxides   106
5.1.1   Silicon dioxide   106

5.1.2 Measurements of silicon dioxide compressibility 108
5.1.3 Dynamic compression of porous silicon dioxide 112
5.1.4 Double compression of quartzite 119
5.1.5 Compression of rutile, cassiterite, magnetite, perovskite, and periclase 122
5.2 Oxysalts 124
5.3 Halides and sulphides 125

6 Rocks 130
6.1 Calculation of rock parameters 134

7 Compression of organic solids 137

8 Liquids 142
8.1 Shock compression of water and aqueous solutions of salts 143
8.2 Saturated hydrocarbons (alkanes) 147
8.3 Nonsaturated hydrocarbons–alkynes and alkenes 148
8.4 Aromatic hydrocarbons 149
8.5 Acids and anhydrides 150
8.6 Alcohols 150
8.7 Ketones 151

9 Conclusion 154

References 157

Index 167

Preface

This book summarizes the results of the fifty-years work of a team of scientists and engineers of the Russian Federal Nuclear Center or All-Russia Research Institute of Experimental Physics (RFNC–VNIIEF) in Sarov (former Arzamas-16), N. Novgorod region. Their effort was concentrated on measuring parameters of compression of condensed materials under intense shock. For the last forty years, the author has participated in this research.

It has been no secret for a long time that the research on shock compression initiated in the USA during World War II and in Russia (then USSR) immediately after the war was directly related to the extremely difficult problem of designing nuclear weapons. No wonder that these investigations remained top secret for a long time, and for ten years after the start of this research only some measurements of shock compression of 'non-secret' metals were published (data about radioactive materials have not been published to this day).

The application of new techniques led to a considerable jump in pressure as compared to the parameters of high-pressure static experiments of that time, $P = 10$ GPa, performed by P. Bridgeman, who was later awarded the Nobel Prize in physics. The first American publications about shock compression at pressures of up to 50 GPa date to 1955, and the first Russian publications at 400 GPa to 1958.

To begin with, the compression parameters of metals, especially those used in structural components of nuclear bombs, were measured. In due time the range of materials studied under shock compression widened rapidly. It included other metals and nonmetallic chemical elements, metal alloys, various mixtures of materials, minerals, organic compounds, liquids, etc.

At present the number of materials studied under shock pressure is several hundred. This research is no longer driven by the needs of the nuclear weapons industry, and its results are more widely used in various fields of science and technology.

In writing this book, I have made an attempt to consider, from different viewpoints, the most important topic of shock compression of matter, namely the general laws which control the compression of condensed matter.

The compression of a material is, naturally, a function of the shock intensity, i.e., the pressure produced by a shock generator. Various shock generators built at the Russian Nuclear Center and driven by high explosives are used to compress studied materials to pressures higher than 2500 GPa, which is a factor of more than three higher than in research centers of other countries.

Our advantage in experiments performed during underground nuclear explosions is even more significant since the maximum pressure achieved by Russian researchers was 10000 GPa, whereas in the USA it was 2000 GPa. In this book attention is focused on measurements at high and superhigh pressures, in which the achievements of Russian scientists are indisputable.

The reader will find in the book a brief review of the history of development of dynamic techniques in Russia, a description of detection techniques, a lot of experimental data concerning high compression of various materials in both laboratory and underground nuclear experiments. The reader will be introduced to the interpretation of experimental data, to general laws which govern the shock compression of various materials, will get an idea of the present state of dynamic high-pressure research in Russia and of the problems which remain unresolved.

The book may be useful for a wide range of readers–both specialists engaged in the high-pressure research and those working in related fields, such as geophysics, planetary astronomy, physics of solids, liquids, and gases. It may also be useful to nonspecialists who are interested in developments and state of the art in contemporary high-pressure physics, specifically, its new division, namely the physics of high-energy density, which was generated and driven largely by progress in the field of intense shock waves.

Many groups of researchers and several eminent scientists made considerable contributions to the development of dynamic techniques in Russia. A significant role was played by Academician Ya.B. Zel'dovich and Prof. L.V. Al'tshuler. The progress in dynamic high-pressure physics was largely due to their personal efforts and work of their colleagues and disciples.

The author has had the privilege of working at the Russian Nuclear Center, where academicians Yu.B. Khariton, A.D. Sakharov, E.I. Zababakhin, and Ya.B. Zel'dovich had created an atmosphere greatly stimulating scientific research, and our results and achievements described in this book would be impossible without the influence of these remarkable leaders.

I am grateful to my wife Nina and our sons Mikhail, Ivan, and Aleksei for help in writing this book and preparing it for print.

1
Introduction

The all-out effort undertaken to produce weapons of nuclear deterrence that was started in the USSR after World War II demanded measurements of properties of various materials, including their shock compressibility, under high pressure. This research was prompted primarily by the necessity of rapidly closing the gap between the USA and USSR in nuclear military technology.

Measurements of material shock compression had been started in 1946, when research aimed at deriving equations of state of various materials was initiated as part of the nuclear program. These equations, which establish relations between thermodynamic parameters, such as energy (or shock compression temperature), and pressure or density, were needed to complete the equation system describing the compression of compressed continuous media.

The equations of state are based on experimental data on the shock compression of materials, in which pressure is determined as a function of density and energy called a shock adiabat (sometimes this relationship is named a Hugoniot equation of state). Unlike the USA, more attention in the USSR, especially in the first years of the nuclear program, was focused on the techniques and devices for studying equations of state of materials. The problem of primary importance was measuring parameters of explosives which triggered the nuclear reaction and metals from which the bomb structure was made. The pioneering research in this field was performed by L.V. Al'tshuler, E.I. Zababakhin, Ya.B. Zel'dovich and many other workers of the Russian Nuclear Center (formerly All-Russia Research Institute of Experimental Physics in Arzamas-16, presently Sarov in Nizhnii Novgorod region).

1

As early as 1947, the shock compressibility of iron and uranium had been studied under pressures ranging up to 40 and 50 GPa (400 and 500 kbar), respectively. But experiments at even higher pressures were necessary for the nuclear program. New measuring devices designed to operate at higher pressures were to be developed rapidly, and this problem was solved very soon. Using these facilities, the pressure was raised to 350 GPa in only one year. Even in the 1990s, such high pressures are available only in three or four countries possessing nuclear weapons, to say nothing of the 1940s.

In 1952 the upper limit of pressure available in Russia for studies of heavy metals was $\simeq$ 1000 GPa. But experimental data were hard to interpret adequately. Therefore, in order to rectify these measurements, new devices were designed early in the 1960s, which had a significantly higher symmetry as compared to previously designed systems, and their impactors moved quasi-isentropically with a very low preheat.

The new facilities were used to correct previous measurements and investigate the compressibilities of many metals, various chemical compounds, ionic and covalent crystals, rocks etc. The upper pressure limit for metals was $\simeq$ 1000 GPa. Given the better reproducibility of the parameters in the new facilities, new data were more reliable and later confirmed by many researchers both in Russia and abroad. Unfortunately, the upper pressure limit in laboratories of other countries was no higher than 500–700 GPa. Therefore the data obtained at very high pressure were verified only in Russian laboratories using facilities of different designs. In particular, in one experiment performed in 1958 in the high-speed mode the compressibilities of iron and uranium were measured at pressures of up to 1400–1800 GPa. A further increase in the pressure produced in laboratory conditions is very problematic, primarily because highly symmetrical, high-speed, and low-preheat impactors are needed.

Most of these difficulties have now been overcome, and devices generating a pressure of up to $\simeq$ 20 Mbar (2000 GPa) in iron, and $\simeq$ 25 Mbar in tungsten and tantalum have been constructed. Apparently, this is close to the limit for systems deriving their energy from chemical explosives.

What techniques are used to measure the pressure and density in compressed samples of materials, and how is the strain measured in shock-wave experiments?

It seems that the most remarkable feature of dynamic techniques is that the strain parameters are derived from kinematic shock parameters, namely the shock velocity D and the velocity of compressed material behind the shock front (or particle velocity) U.

Thermodynamic parameters, such as pressure P, density ρ, and shock compression energy E are expressed as functions of D and U using the conservation equations of momentum $(P - P_0 = \rho_0 DU)$, mass $(\rho = \rho_0 D/(D - U))$, and energy $(E - E_0 = 0.5P(\rho - \rho_0)/\rho\rho_0)$. As concerns the temperature T, which is another important thermodynamic parameter, no techniques for its direct measurement (except relative measurements in transparent media) are available at present. Therefore the temperature of shock compressed materials is usually calculated using equations of state, which are based on more or less reliable models of condensed materials.

The equations given above indicate that dynamic high-pressure experiments do not demand state-of-the-art facilities with diamond anvils and their parameters are not limited by the strength of construction materials, i.e., the technique allows one to experiment at superhigh pressures not available in facilities with pressurized chambers.

On the other hand, the shock compression of materials, unlike static pressure, leads to a considerable increase in temperature, and the higher the pressure acting on a studied material, the higher the overheat temperature. This means that an equal density of a compressed material is achieved in a shock experiment at a higher pressure than in a static experiment. The two processes are compared in the $P - V$ diagram $(V = \rho^{-1})$ given in Fig. 1.1. The lower curve is a static isotherm of 'cold' $(T = 0)$ compression, and the upper curve is a shock (Hugoniot) curve. The equation $E_H - E_0 = 0.5P_H(V_0 - V)$ yields an increase in the intrinsic energy of a material due to compression. It contains the 'cold' fraction (it is equal to the area limited by the curve $P_c(V)$ and the ordinate axis) and the thermal fraction, $E_H - E_c$ (hatched area in the diagram). By extrapolating the curves of P_c and P_H to smaller volumes, one can easily see that the thermal component increases with the pressure, hence the temperature also grows. As a result, there is a so-called shock compression limit, beyond which a material cannot be compressed by a shock, regardless of the resulting pressure. The limiting compression usually corresponds to a five to six-fold increase in density, and the typical pressure at such

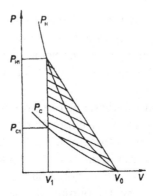

Fig. 1.1. $P - V$ diagram of shock compression: P_H is a Hugoniot adiabat; P_c is an isothermal cold compression curve $(T = 0 \text{ K})$.

compressions is several hundred terapascals. In other words, the relative compression of a solid material cannot be higher than five to six-fold at arbitrarily high pressures.

This feature of the dynamic compression should not be considered as its flaw. From the viewpoint of equation of state studies, both the static and dynamic techniques complement each other. In particular, by comparing both types of measurements, one can estimate thermal contributions to any thermodynamic parameters in processes when the temperature changes with compression. Dynamic measurements are very significant for testing theoretical models of materials, for calibrating strain sensors operating at very high pressures, for studies of first-order phase transitions in materials at high pressures, and for other scientific and technological applications.

The dynamic compression technique has been extensively used both in Russia and other countries. In our experiments, we have mostly used facilities with high explosives generating record pressures of up to 25 Mbar.

The progress in our high-pressure techniques is illustrated by Fig. 1.2, which shows the maximum available pressure versus time. The dashed line shows preliminary compression parameters recorded in 1952. The final data are plotted by the solid line. The diagram indicates that the drive to higher pressure was not easy since it took almost fifty years to achieve the record pressure of 25 Mbar.

Now let us briefly review the techniques of shock compression

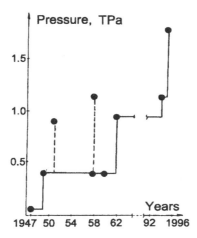

Fig. 1.2. Maximum available pressure in laboratory experiments with iron versus date: preliminary data (dashed line); final data (solid line).

measurements, i.e., measurements of the shock (D) and particle (U) velocities. In Russia the two basic methods are the deceleration [1] and impedance-matching techniques [2]. Both methods are based on strictly defined principles and are absolute in the sense that the thermodynamic parameters which characterize the compressibility are uniquely determined by the parameters recorded in the experiment, namely the shock D and particle U velocities, and conservation equations.

In the deceleration technique, the velocity of a moving plate hitting a resting target made from the same material and the shock velocity in the target are measured concurrently. At the moment of impact, the velocity and pressure are continuous across the impactor–target boundary owing to the continuity conditions and equality between action and counteraction. It is easy to prove that if W is the impactor velocity (measured in the experiment) and U is the velocity of the interface, which is equal to the particle velocity in the target, then $U = W/2$.

The shock velocity D, as well as W, is determined by measuring the shock transit time across the sample. It is detected by sensors of any type reacting to the pressure jump across the shock front. Using the conservation equations, the shock compression thermodynamic parameters are derived from measurements of D and W. By varying the impactor velocity, one can study various compressed states in the tested material, i.e., determine its parameters in appropriate ranges of pressure and density.

The only source of uncertainty in this technique is, in effect, the impactor preshock heating, which is a function of the intensity of the incident detonation wave. Generally speaking, corrections for the preheat derived using the equation of state of the impactor material should be introduced. The corrections are designed to approximately describe all the heated states of the impactor in terms of the 'cold' initial state. This effect, however, had been taken into account in designing accelerating systems at the Russian Federal Nuclear Center (see below), therefore the preheats of impactors made, for example, from the St3 low-doped steel and moving at a velocity of 1 to 20 km/s were approximately constant and not higher than the preheat of steel exposed to a normally incident detonation wave.

The second widely used method is the impedance-matching technique. It can be employed only if the equation of state or, at least, the Hugoniot of the shield material downstream of which tested samples are placed is known (the Hugoniot of the shield is usually determined using the deceleration technique). In this case the sought state of the tested material is determined by the intersection point between the straight line defined by the equation $P = \rho_0 DU$ and the expansion curve (or the double compression Hugoniot) of the reference material (shield) originating from states corresponding to its parameters under the incident wave. In practice the so-called 'mirror approximation' is used, when the expansion (or double compression) curve is substituted with the Hugoniot of the standard material. It was confirmed both by calculations using various equations of state and experimentally [3] that this substitution does not lead to a considerable loss of accuracy if the pressure in the initial state of the shield and in the tested material differ by a factor of less than five. Using different measuring devices appropriate to different velocities W and, hence, different shield parameters, we obtain a set of states of the tested material, which define the position of its Hugoniot (the curve of shock compression). In the data processing technique employed in our laboratory, each value of the shock velocity is an average of five to eight independent measurements. An equal number of impactor velocity W measurements are usually accumulated.

To conclude our discussion of laboratory shock compression techniques, note that the accuracy of state-of-the-art measuring devices, the possibility of improving their designs, and the uniformity of energy density generated by shocks in specimens about one millimeter thick indicate that other sources of energy, such as lasers, electromagnetic generators, various electromagnetic and electrodynamic accelerators, multistage gas guns etc., will be hardly able to compete with facilities using explosives for the foreseeable future.

There is, however, another source of energy which offered new prospects for researchers. This source is nuclear explosion. The pressures which can be generated in tested materials in such experiments may be not only comparable, but exceed by tens and even thousands of times the highest values achieved in facilities using high explosives. It is quite a common feature of a large-scale scientific program that progress in the research started to measure parameters needed to develop nuclear charges was in due time driven by the energy generated in their explosions.

After the ban on nuclear explosions in space, ocean, and atmosphere imposed by the Moscow Treaty of 1963, the only medium where tests of nuclear charges were permissible was rock masses. It would have been an inexcusable mistake not to use in scientific research the enormous energy of a nuclear explosion generating intense shock waves in rocks surrounding the site, and, fortunately, the researchers did not make it.

The first attempt to measure compressibility using the energy of an underground nuclear explosion undertaken by workers of the Russian Nuclear Center in Arzamas-16 dates to late 1965. It was undertaken during a test at the Semipalatinsk testing ground. The aim was to measure the rock compression under a pressure of 50–100 GPa and a relative compressibility of the Pb–Fe systems under a pressure $P > 1$ TPa. For technical reasons the experiment failed. The first successful measurements of the shock compressibility of a rock (granite) were performed by workers of Chelyabinsk Nuclear Center in the first half of the following year. The data obtained under a pressure of about 350 GPa were in agreement with laboratory data of the Arzamas Center at approximately equal pressures, which were available at the time.

Now let us briefly discuss the main trends in underground test research. They are controlled by the features and possibilities offered by these tests, namely the wide range of pressure in the experiments (depending on the explosion energy and distance between the site of explosion and tested sample), the geometry of shock propagation, which is one-dimensional in most cases, the possibility of testing samples whose dimensions are several orders of magnitude larger than those of laboratory samples, and several other conditions.

Thus the following main directions of research were defined:

1. Verification of laboratory data at a considerably longer time of shock action on a sample (so-called scaling factor). In particular, this increase in the time scale stimulated the first measurements of rock compressibility in 1965–66.

2. Measurements of compressibility of condensed matter in the intermediate pressure range between pressures available in laboratory experiments ($P \leq 1\,\mathrm{TPa}$) and those at which a reliable theory had been developed (tens of terapascals). The aim of these measurements was to plot reliable interpolation curves in this pressure range and to select a theoretical model which was consistent with the experimental data.

3. The selection of the theoretical model of matter most adequate to fit experimental data.

Generally speaking, these goals demand measurements which are absolute in technical terms, but before using these techniques under the specific conditions of a nuclear explosion (they will be described below), we decided to measure relative compressibilities of materials. This technique, which had been well developed by that time, demands measurements of the velocity of a shock propagating consecutively across layers of two or three tested materials. Usually the upstream sample is a standard one whose equation of state (or at least its Hugoniot) is assumed to be known throughout the pressure range needed for further data processing. The shock velocity in the reference sample determines through its equations of state the parameters (P, U) which are the initial data for calculations of compression parameters of tested materials in the impedance matching technique.

Until 1977, i.e., for more than ten years of experiments performed during underground tests (at a pressure of more than 1 TPa), the technique of relative compression measurements was used. In Russia the shielding (standard) materials were in most cases iron (sometimes lead, copper, or uranium) and aluminum, in the USA molybdenum and aluminum. The selection of these standard materials in Russia was primarily determined by the large atomic numbers Z of these metals (except aluminum). Quantum statistical techniques for calculating Hugoniots of these materials were becoming more reliable at the time (let us recall that those experiments were performed in the 1970s, when the only statistical model of matter under a superhigh pressure was the Thomas–Fermi model for electronic states). Therefore we could plot more dependable interpolating curves using these models in order to establish the relationship between experimental data in the studied region of the P–ρ diagram and calculations. As for aluminum, whose atomic number is relatively small, it was used only because no other metal which could be employed as a standard for low-density materials was available. Later the absolute compressibilities of iron and aluminum were determined throughout the required pressure range, thus all the relative measurements could be interpreted in terms of absolute parameters.

Progress in nuclear experiments with shock compressed materials starting from the first measurements of 1965–66 mentioned above and until 1977 was totally controlled by the Federal Nuclear Center at Arzamas-16. At the end of 1966, in one experiment on the island of Novaya Zemlya, a pressure higher than 1 TPa was produced. Some results of that experiment were given in Refs. [4–8]. The compressibility of the Fe–Pb–U system at pressures of about 3.1 (Fe), 3.4 (Pb), and 4.0 TPa (U) was studied [4], and the system of lighter materials (aluminum–quartzite SiO_2) was tested at a pressure of up to 2 TPa [5]. I recall one occasion related to measurements of quartzite compressibility in this experiment under a pressure of about 0.7 TPa.

By force of circumstances, we had the possibility of another measurement in addition to planned experiments. But we had neither facilities nor materials to manufacture a sample for this experiment at the testing ground. When I turned over in my mind the materials we wished to investigate, it occurred to me that it might be possible to find a sample of quartzite with the required thickness of about 100 mm and area (the inscribed circle diameter was to be no less than 300 mm) in the rubbish heap near the tunnel. We had more than enough slabs of quartzite because a large section of the tunnel was drilled through this rock. But in addition to large dimensions, the desired sample of rock was to satisfy another condition, namely, at least one of its sides should be perfectly plane to match an aluminum shield. Can you imagine that nature had fabricated such a piece of rock! After a long search we found one and performed measurements of its compressibility. An experiment under a pressure of about 0.7 TPa with such a relatively light material as SiO_2 was a remarkable achievement in the 1960s, in the early stage of our effort.

The relative compressibilities of metals in the Pb–Cu–Cd system were measured in 1968 under a pressure of 1.5 TPa [6], then this system was tested in 1970 under a pressure of $\simeq$ 5 TPa [7]. In that year the upper pressure limit for the Fe–Pb couple tested previously was also increased to 5.2–5.8 TPa.

At the same time, the data on the compression of light materials, such as water (1.4 TPa) [8], quartz of various initial densities ($P \simeq$ 1 TPa) [9], and Plexiglass $(C_5H_8O_2)_n$ ($P \simeq$ 0.6 TPa) [9] with aluminum used as a standard, and absolute laboratory measurements at a relatively small pressure (i.e., in the range where absolute measurements of aluminum were available) had been obtained. Those experiments yielded data about the compressibility of graphite [10], rutile, rock salt [11], several rocks [12], and the compressibilities of aluminum (at $P \simeq$ 0.5 TPa with an iron shield) [9] and several other (including porous) materials were verified. In a series of experiments in 1971–75 relative compressibilities of porous materials, including copper, tungsten, iron, and some heavy compounds were measured under a pressure of 1–2 TPa [13].

Since the early 1970s several attempts have been made to determine the compressibilities of the standard metals using absolute measurement techniques. At the Federal Nuclear Center some of these measurements used the deceleration technique, i.e., the velocities of the impactor and of the shock in targets made from the same material were determined. In 'full-scale' experiments* such measurements proved to be so complicated that, in terms of accuracy and easy interpretation, those measurements were beyond admissible margins. The major problems were to design 'zero-preheat' plate accelerating systems providing highly symmetrical plate propagation, to select sensors not susceptible to the shock propagating in air ahead of a moving impactor, and some other technical challenges. Nonetheless, several experiments were performed successfully, and the data of 1970–74 were published in 1992–93 [11, 14, 15]. The shock pressure obtained in iron in these experiments was 4.0, 5.5, and 10.5 TPa. The results of these studies will be discussed in detail below. At this point, let us note that,

---

* In this book we shall apply this term to underground nuclear tests.

given the Hugoniot of iron determined under pressures ranging to these values by absolute methods, we could express all the data on the relative compressibility of metals which were discussed above in absolute terms.

By the beginning of 1980, the upper pressure limit in relative measurements of the system of Fe (reference)–Cu, Pb, Ti (tested materials) had been raised to 20 TPa [16]. The uncertainties of some measurements obtained in those experiments were two times as large as those we could tolerate at the time. This was the reason why we did not publish those data, since we hoped to rectify the results using other methods, but because of circumstances beyond our control, we could not do so.

In 1980 investigations of the relative compressibility of Pb–SiO_2–Al–H_2O (tested samples were placed downstream of an iron shield) were published by the group from the Chelyabinsk Russian Nuclear Center [17]. That publication presented for the first time measurements performed beyond 10 TPa. Although the starting data in that publication were incomplete and the analysis revealed considerable uncertainties in some measured parameters (specifically in those referring to lead, see below), it was the first item in a new series of papers describing measurements under record compressions in underground nuclear explosions.

The same group performed measurements in 1985 under an enormous pressure of 700 TPa [18]. In that experiment, they obtained a large set of self-consistent data about the systems of Fe–Al, Fe–Pb, Pb–Fe, Fe–H_2O, and Fe–SiO_2. But in discussing these data, one note is relevant: thermodynamic parameters derived from measurements of Fe–SiO_2, Fe–H_2O, and Fe–Al systems under such high pressures as those achieved in those experiments and under comparable pressure differences between Hugoniots of iron (shield) and tested materials depend on the assumed equation of state of iron used to calculate expansion Hugoniots, which are necessary to determine the compression parameters of lighter materials. The uncertainties in positions of the Hugoniots lead to an additional error in the particle velocity. This, naturally, does not concern systems of metals with close dynamic compressibilities, such as Fe–Pb and Pb–Fe.

An important step in the development of new techniques was the suggestion by V.A. Simonenko and L.P. Volkov (Chelyabinsk Nuclear Center) to determine the absolute compressibilities of materials using the so-called γ-reference . They applied the technique to measurements of aluminum [19, 20]. In the experiment they detected the motion of γ-active pellets which were imbedded into a tested sample and activated by the radiation due to the nuclear explosion. The motion of the pellets was detected by counters placed behind collimating slits converting γ-quanta to optical pulses. It was taken for granted that the velocity of references (europium dioxide, whose initial density is equal to that of aluminum) was equal to the aluminum particle velocity. The γ-reference technique was used in several measurements of aluminum compressibility under terapascal pressures (measurements by Chelyabinsk Center dating to mid-1970s [20] and by Arzamas Center in 1978 [21]) which were a factor of 1.5–2 larger than the highest pressure produced in the laboratory.

In order to complete our description, let us recall the milestones in the history of nuclear-explosion compression research in other countries.

In the mid-1970s (the publication dates to 1978) Ragan III (Los Alamos National Laboratory) suggested a new technique for measuring absolute compressibility and applied it to molybdenum [22]. The method is based on detecting two kinematic parameters of a shock propagating in a material, namely the shock velocity derived from the time of light flashes due to its arrival on the molybdenum surface and the particle velocity calculated using the shift of neutron resonances in moving nuclei with respect to those at rest (the Doppler effect). The high pressure was due to the fission energy of uranium-235 nuclei in the base plate triggered by neutrons from a nuclear explosion. Unfortunately, because of uncertainties due to the neutron pulse width and spread of resonance peaks, the particle velocity was determined only to within 5%, which was not satisfactory for single-shot calibration measurements. Nonetheless, we should do justice to the imagination of the author, who used for the first time a feature of the nuclear explosion, namely the intense flow of neutrons, to record hydrodynamic characteristics of compressed matter.

In subsequent nuclear explosion experiments, Ragan III and coworkers [23–25] obtained data on the relative compressibility of metals and compounds under conditions which were quite similar to those of our experiments (a pressure range of $P < 7$ TPa).

Finally let us recall the first publication in 1983 [26] by the Livermore National Laboratory about their research on metal compressibility. The publication by Chinese researchers about iron compressibility dates to 1990 [27].

Note that the pressure range in experiments carried out in other countries was more narrow than in measurements performed by scientists of Russian nuclear centers.

The diagram in Fig. 1.3 shows maximum pressures at which the shock compressibility of iron was measured by absolute techniques in nuclear experiments plotted against date. The maximum pressure is 10 TPa. In our experiments the standards were other materials, such as aluminum, copper, lead, and metal alloys. Their Hugoniots had also been measured by absolute techniques throughout relevant pressure ranges. We should stress that, like laboratory measurements, nuclear experiments were labor consuming and demanded a lot of effort. In particular, the specific conditions of underground tests, in which we used bulky samples, specially designed detectors and electric circuits, demanded that cable lines be protected from exposure to high-speed jets and other effects, so we had to develop a strict routine for these experiments.

Not all components of this routine were predetermined and based on general rules, some of them derived from our experience. In the first nuclear test including shock compression measurements (1965, Semipalatinsk) we used NB-11 oscillographic recorders to measure time intervals. In these recorders timing diagrams were recorded using a mechanical scanning device, which drew a photographic film in front of a cathode-ray tube. At that time these

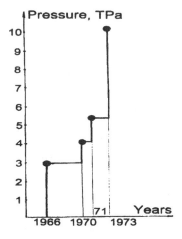

Fig. 1.3. Maximum available pressure in experiments with iron during underground nuclear tests.

were reasonably fast recorders.

All preparatory tests passed smoothly. We had performed all the necessary preliminary routines with the apparatus, recorded testing signals, and adjusted the film-winding devices. In the night before the test there was a slight frost, the first one of that fall. We did not give much attention to this event and neglected the fact that the rack with those devices was not defrosted. The lubricant in the friction units of the film winding devices, naturally, froze, the film could not be accelerated to the desired speed, and after an initial abrupt pull it was torn apart in several devices. Of course, we made appropriate conclusions after that experiment. The data were saved by fall-back recorders, in which signals were recorded on immobile films.

We ran into another unexpected problem in preparing this experiment. We fixed bulky anchor bolts on ground surfaces of so-called measuring planes fabricated in the rock to support rather heavy (more than 200 kg) aluminum shields for measurements of rock compressibility. The aluminum blocks had bolt holes that were designed to fit to positions of the anchor bolts. Alas, these holes did not match up with the bolts. And they could not match because it was impossible to position the bolts in the rock as accurately as holes were machined in the aluminum slabs (the holes for the bolts were made by a pneumatic drill). So we had to fit the aluminum shields to the anchor bolts using hack-saws. You can estimate the scale of this work given that each shield was about 40 mm thick and some cuts made by hack-saws were more than 70 mm deep. We had to make manually two or three cuts in several shields! We learned this lesson well, and later holes for anchor bolts were drilled in the rocks when the aluminum shields were at the site to match up bolt positions to the holes machined in a shield.

Here is the last example (but not our last case in practice!) of unexpected problems encountered in underground experiments. When analyzing results of the first experiments, we found that a fraction of information was lost, tentatively because of ruptures in coaxial cables due to cumulative jets generated when the shock arrived on the rough rock surface (tunnels were dug using explosives). In order to prevent losing data, we had to shield the experimental apparatus and cable lines conducting signals or, plainly speaking, to bury them in a thick layer of sand or gravel.

After these and other failures in the first experiments caused by the design of experimental apparatus and measuring techniques, we decided to plan the preparatory operations performed in the tunnel and with measuring devices more carefully.

By the 1990s, researchers in Russia and other countries had measured the shock compression parameters of a wide range of materials. One can hardly find a class of compounds, a notable group of elements, mineral structures, or organic compounds, etc., whose dynamic compression has not been studied. Therefore it is impossible to give an exhaustive description of experiments with all the materials in one review. Thus we shall discuss the properties of typical representatives of studied classes of materials and focus our attention on their shock compression. This refers mostly to laboratory data, which constitute the bulk of the available information. The few measurements performed during underground nuclear tests are in most cases unique pieces of information and should be discussed separately.

As we noted above, the first tested materials were metals. Later research on their properties remained a topical problem for several reasons, in particular, because this class of materials is the most simple but, at the same time, very versatile, and the problem of modelling their behavior under dynamic loads is relatively simple, but, on the other hand, very important and advisable in view of studying other materials.

Most of this review is dedicated to metals, in accordance with the importance attached to their dynamic properties throughout the period after World War II.

In addition to normal ambient conditions, we shall discuss investigations of the shock compression of porous metals, i.e., metals with a lower initial density; of melted and overcooled metals (single measurements); of metal alloys; hydrides, carbides, and nitrides of metals; halogenides; oxides; oxygen salts; some minerals and rocks; and organic compounds.

In the last section, we shall consider investigations of liquids. It presents studies of water and snow, aqueous solutions of some salts and a large group of organic compounds, such as alcohols, anhydrides, acids, etc.

Experimental data are presented in the form of graphs and diagrams. Most of the discussed results have been obtained at the Russian Nuclear Center in Sarov (Arzamas-16). Data from other laboratories are used as illustrations.

2

Measuring kinematic parameters of shock waves

2.1 Laboratory shock generators

In laboratory shock generators, thin metal impactors of plane or spherical configurations are accelerated by explosion products from high explosives. These generators were used to determine compression parameters of dozens of various materials, including iron and aluminum, which are basic standard materials in the impedance-matching technique.

Let us first discuss 'plane' generators of shock waves, i.e., those in which plane impactor plates are accelerated by explosion products. The main requirements imposed on such devices are the following:

1. Good symmetry of the flying impactor; on its arrival at the target, the spread of the impact time of its different parts must be within $5 \cdot 10^{-8}$ s.

2. The impactor plate must not disintegrate during the flight.

3. The shock parameters in a tested sample must be constant throughout the measurement time.

In experiments under pressures of 30 to 50 GPa, we used devices (Fig. 2.1, Ref. [28], pp. 28, 60, 225) including a charge of explosive (depending on the required pressure, the charge height and its composition are varied) with a specific correcting lens to generate a detonation wave with a planar front. After the detonation wave has reached the charge end, explosion products propagate across a 3–5 mm air gap, hit a metal driver, and generate in it an almost stationary shock wave, i.e., a shock with a pressure approximately constant with time.

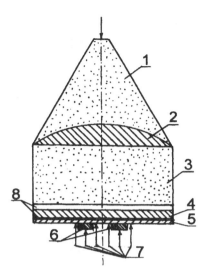

Fig. 2.1. Direct contact measuring device: (1) lens-shaped high-explosive charge; (2) correcting lens; (3) main charge; (4) impactor plate; (5) shield from a standard material; (6) tested sample; (7) shorting-pin detectors; (8) air gaps.

Usually tested samples are either solid disks with a diameter of 40 to 60 mm or cylindrical slabs with a diameter of 10 to 15 mm. In the latter case, a group of samples of different materials are usually placed downstream of the driver. The sample thickness in experiments is varied between 3 and 6 mm.

The velocities of impactors and shocks in tested samples are usually determined using pin detectors coated with insulating enamel. When a shock front or a moving plate hits a detector, the gap between oppositely charged electrodes is closed, and the appropriate signal is recorded by an oscilloscope or a digital device with a time resolution of $\pm 5 \cdot 10^{-9}$ s. Groups of twelve to sixteen detectors are placed on both sides of tested samples (Fig. 2.1) to determine the time of the shock transit across the sample. Readings from different sensors indicate the symmetry of the impactor or shock propagation (its deviation from the planar shape). When necessary, these multiple measurements are used in processing experimental data.

At higher pressures, aluminum or iron plates accelerated by explosion products to a velocity of 3.5 to 6.5 km/s are used (Figs. 2.2, 2.3, Ref. [28], p. 28). Aluminum impactors in these devices are disks 2 to 6 mm thick and 60 to 70 mm in diameter pressed into steel disks of equal thickness (Fig. 2.2). A detonation wave arriving on the metal surface generates in the steel disk a pressure higher than in aluminum, thus an additional thrust acts on the outer area of the aluminum disk to flatten the flying impactor plate.

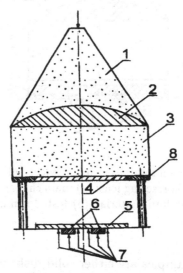

Fig. 2.2. Diagram of the facility with an accelerated aluminum impactor: (1–7) the same as in Fig. 2.1; (8) peripheral steel ring.

Steel impactors (Fig. 2.3) are disks 65 to 70 mm in diameter, 1.1, 1.5, or 2.2 mm thick, and their peripheral area is 5 mm thick. The function of the thicker part, as in the previous case, is to flatten the flying impactor plate. Under these conditions, the shock pressure generated in the tested sample is also constant with time.

Higher shock parameters are produced by generators of a spherical configuration. The structures of such devices are shown in Figs. 2.4 and 2.5. The first diagram shows a hemispherical device which contains a high-explosive charge, whose detonation is triggered simultaneously on the entire outer surface, with a metal (low-doped St3 steel) driver shell. This shell is accelerated by the explosion products and converges towards the center of the device. Its velocity increases as it moves to the center, therefore the pressure in a target may be varied by varying its distance from the center. In this specific case, you see a diagram of the device which was used in the 1950s to measure the compressibility of met-

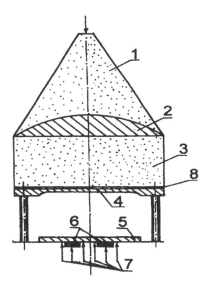

Fig. 2.3. Facility with an accelerated steel impactor: (1–7) the same as in Fig. 2.1; (8) 1-mm Plexiglass spacer.

als under pressures of up to 400–500 GPa [1, 2]. Unfortunately, this configuration of the device had some flaws which made the interpretation of experimental data more difficult, such as:

– rather low symmetry of the flying impactor (hence poor uniformity of the shock in the target), therefore the sample thickness

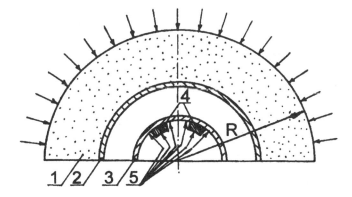

Fig. 2.4. Facility with overcompressed detonation wave: (1) high-explosive charge, whose detonation is triggered on the surface; (2) steel impactor shell; (3) iron shield; (4) tested samples; (5) shorting-pin detectors.

was increased to give a higher accuracy of measurements and corrections for the 'shock convergence' were introduced;

– relatively high preheat of the shell due to the converging detonation wave (in some cases up to 1500^0 C). Therefore corrections had to be introduced in processing experimental data.

The first correction took into account the difference between the variable shock velocity D_{exp} in a sample of a finite thickness and the instantaneous shock velocity D_{inst} that would be detected in an infinitely thin sample placed at the same site and hit by the impactor. This correction is a function of the sample thickness and in experiments with some metals [1, 2] it was 2–3% of D_{exp}.

The impactor preheat was taken into account by introducing a negative correction to its velocity (a transition to the 'cold' initial state of the impactor). In the experiment described in Ref. [2] this correction was within 1% of W_{exp}.

A diagram of a modified measuring device is shown in Fig. 2.5. Its design was changed in order to get rid, as far as possible, of the

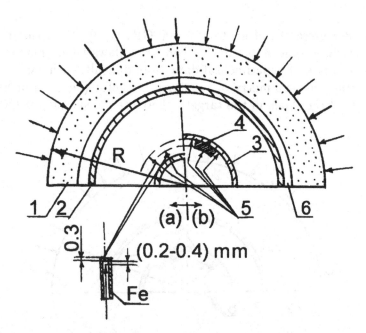

Fig. 2.5. Modernized facility with a low-preheat impactor shell: (1–5) the same as in Fig. 2.4; (6) air gap. The left-hand half of the diagram (a) shows the configuration for measuring the shell flight velocity; the right-hand half (b) shows the layout for measuring the shock velocity.

flaws of previous devices. This was done, primarily, by improving the symmetry of the converging impactor shell and shock propagating in the target. As a result, the thicknesses of the shield and tested sample could be reduced without a loss in the accuracy of measured parameters, hence the correction for the shock convergence could be also reduced. Given the higher symmetry of the device, the measurement reproducibility also improved.

The next step was to reduce the shell preheat. This was done by introducing a thin air gap between the shell and the explosive. This method of reducing the pressure and, hence, temperature was proposed in 1948 by E.I. Zababakhin and Ya.B. Zel'dovich. It was widely used (Ref. [28], p. 60) both in devices with accelerated plane impactors and in modified facilities of spherical configuration [29]. In both cases the method demonstrated its efficiency. The combination of these measures lead to the design of a new type of explosive shock generators with different positions of targets with respect to the center, shell thicknesses, and air gaps between shells and explosives.

Velocities of an impactor and shock in a tested sample are measured, as in planar devices, using coaxial electric contactors with electric pins coated with insulating enamel. When a shock or an impactor hits a contactor, an electric breakdown between oppositely charged electrodes takes place and is detected by a fast oscilloscope or a digital device with a time resolution of less than $5 \cdot 10^{-9}$ s. Groups of contactors are placed on both faces of tested samples to detect the shock transit time (or the time of the impactor flight over a given distance). The symmetry (or flatness) of the impactor motion and shock propagation is also checked using contactor signals. When necessary, the asymmetry is taken into account in processing experimental data. The kinematic parameters, such as the shell flight velocity W_{sh} and shock velocity D, are measured at several distances from the device center. This allows us to plot the reference iron Hugoniot in a pressure range of 280 to 900 GPa. Such measurements determined the parameters of impedance-matching experiments with other materials when iron was used as a standard. The impedance-matching technique was used in numerous experiments to determine compressibilities of various materials, and the results were reported in many scientific publications (see, for example, Ref. [28], pp. 27, 36, 123, 208, 236, 249).

In the early 1990s, a device was designed at the Federal Nuclear

Center [30, 31] to measure the compressibility of heavy metals ($\rho_0 \geq 15$ g/cm^3) at a pressure of up to 2.5 TPa (the spherical impactor velocity in the detector region $W = 18.6$–23.0 km/s). The device was used to check some results of full-scale experiments (see below).

To conclude this discussion, we should stress that sample dimensions were selected in both laboratory and full-scale experiments so that the diameter should be at least twice as large as the thickness. Then the detectors imbedded in the central zone on the downstream side of the sample are not affected by relief waves propagating from the cylindrical sample surface.

There is another selection rule for thicknesses of impactors, shields, and samples to get rid of the effect of centrally symmetrical relief waves reflected from the interface between the impactor and explosion products (or driver explosion products) on the parameters of the shock front.

2.2   Specific features of full-scale experiments

The impedance-matching technique is the basic method used in underground nuclear tests, at least when the Hugoniot (equation of state) of the shield material is derived either from experimental data or from some generally accepted model.

A diagram of the experimental layout used in nuclear impedance-matching measurements is given in Fig. 2.6. A polished base plane oriented normally to the direction from its center to that of the explosion is formed at a selected distance from the nuclear charge center in the rock mass. One or several slabs of tested materials are placed behind one another. The first slab is typically standard material. If the Hugoniot (or the equation of state in a more general case) of the standard material is known, all sought parameters are derived from measurements of the velocity of the shock propagating across the standard and tested materials (in the impedance-matching method, the boundary problem on the interfaces between contacting materials is considered).

As concerns absolute measurements of characteristics of standard materials in underground tests, we have noted above that those of iron were determined by the deceleration technique, those of molybdenum were derived from Doppler measurements of nuclear resonances, and those of aluminum by the γ-reference

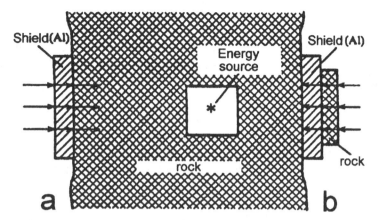

Fig. 2.6. Diagram of full-scale experiment during underground nuclear explosion using both direct (b) and reversed (a) impedance-matching techniques.

method.

Let us discuss some features of shocks generated by underground explosions. The main one is the damping of a divergent shock. The higher the shock amplitude, i.e., the closer the detectors to the explosion zone, the higher the damping decrement. At sufficiently high energy of explosion, the pressure in measuring devices may range from tens or hundreds of terapascals next to the charge to those corresponding to compression waves (fractions or units of gigapascals). Hence this large shock damping should be taken into account in processing experimental data because velocities are derived from measurements over finite thicknesses of either shields or tested materials. Usually shock velocities are compared on interfaces between two samples, and the instantaneous velocities on the boundary are derived from their average values either by taking into account the shock damping or through a more complicated calculation procedure.

There were cases in our experiments when the shock damping could be ignored. All these cases are stipulated in the text.

Another note concerning full-scale experiments. From the viewpoint of dynamic measurements, there are practically no boundaries in detector zones which affect the shock front if the rock mass is uniform. Therefore, the results of these experiments can be processed without taking into account rarefaction waves reflected from boundaries and affecting parameters of the shock front. Therefore thicknesses of samples used in full-scale experiments are not limited. Usually the thicknesses are 10 to 15 cm* (this dimension is largely dictated by the ease of mounting samples and building structures supporting detectors) in the case of man-made samples and 40 to 60 cm if parameters of rocks surrounding the testing site are measured.

In the latter case, it seems more convenient to exchange the positions of samples (in impedance-matching experiments) and to place the tested material (in this case the rock) upstream of the shield from the standard metal. If the positions of Hugoniots of the tested rock and standard material plotted in $P-U$ coordinates are close to each other (the pressure difference between the two curves is within 20–30%), the particle velocity in the tested rock derived from experimental data is practically the same when the deceleration adiabat of the rock is substituted by the shield Hugoniot. Using this approximation called the 'inverse impedance-matching method' [5, 11, 12], we have measured compression parameters of some rocks.

These are the main features of full-scale experiments.

In conclusion, note that shock generators designed at the Federal Nuclear Center for compression measurements are suitable practically for solving any problems confronting researchers working in this field of high-pressure physics.

Specific features of full-scale experiments undertaken to measure the absolute compressibility of the standard metals–iron and aluminum–will be discussed below.

---

* Limitations due to rarefaction waves reflected from side surfaces are, naturally, the same as in laboratory conditions, i.e., the sample diameter at this thickness is 30 to 40 cm.

3
Shock compression of metals

3.1 Absolute laboratory measurements of kinematic parameters

Shock compression of about sixty metals of the periodic system, i.e., of the majority of metals, has been studied over various ranges of pressure.

The reported maximum pressures achieved in laboratory conditions are 1.8 TPa for iron, 1.0 TPa for aluminum, 2.5 TPa for tantalum, 1.3 TPa for titanium, and 2.1 for molybdenum [30, 31]. More than ten metals have been studied under pressures of up to 1.0 TPa, the rest of them, depending on their initial density, have been tested in a range of 60 GPa (lithium) to 500 GPa (tungsten) [32].

Let us recall that the parameters directly measured in dynamic experiments are the shock velocity D and the particle velocity U behind the shock front. The thermodynamic parameters, such as pressure, compression, and energy are derived using conservation equations, and the temperature is derived using the equation of state. We shall analyze most of the experimental data in terms of the directly measured parameters D and U.

A remarkable feature of Hugoniots plotted in $D - U$ coordinates is the linear or close to linear dependence between these parameters in fairly wide ranges of parameters. This linearity between the shock velocity and particle velocity in certain ranges of these parameters has not been proved theoretically, but was observed in hundreds of experiments with various materials ranging from gases to solids in their initial states. Certainly, this statement is an approximation to the real state of things and, to some extent, depends on the measurement accuracy of shock parameters and density of experimental points on the plots. We shall demonstrate below that real experimental curves deviate from straight lines because of physical processes in materials, such as melting, evaporation, compression of electronic orbits, etc. Nonetheless, these deviations are not too great to discard the general tendency, namely the linearity of $D - U$ Hugoniots. A fairly detailed classification of metals with respect to the shapes of their $D - U$ curves was given in Ref. [32]. This classification has undergone only minor modifications to this day, specifically because the number of experiments with metals in the considered pressure range since our publication [32] has been small, therefore we shall adhere to the classification of Hugoniots given in Ref. [32] with some corrections. According to this scheme, Hugoniots of all metals are classified into five groups.

The first group (Fig. 3.1) includes metals whose $D - U$ curves are linear throughout the studied range of wave velocities:

$$D = s_0 + aU,$$

where s_0 is the provisional sound velocity, i.e., the shock velocity D extrapolated to $U = 0$ (in most cases s_0 is close to the real sound velocity in volume), and $a = \mathrm{d}D/\mathrm{d}U = \mathrm{const}$ is the slope of the $D - U$ curve. The adiabats of eleven metals–Li, K, Be, Mg, W, Mo, Re, Ir, Au, Ba, and Ga–are linear throughout the studied range of velocities.

Four of them–K, Mg, W, and Mo–have slopes $a = 1.2$–1.25; in Li and Be $a = 1.1$; and in the rest of them $a \geq 1.35$ (the largest slope $a \simeq 1.56$ is that of Au and Ga Hugoniots).

In the range of superhigh pressures, when the compression is close to its limit ($\sigma = \rho/\rho_0 \geq 4$), the existing theoretical models yield $D'_U = \mathrm{d}D/\mathrm{d}U = 1.2$–$1.3$. You can see that the measured slopes of K, Mg, W, and Mo Hugoniots fall in this range. This leads us to the conclusion that in these metals the $D - U$ curve is linear in a range up to the states which can be described by theoretical models developed for the high-pressure limit with considerable confidence. As concerns other metals of this group, the difference between slopes of their Hugoniots and the theoretical limit indicates that at shock parameters U and D higher than the experimentally studied ranges, their Hugoniots should switch to those calculated in the high-pressure limit ($a \simeq 1.25$). These

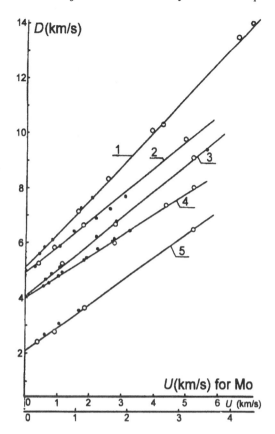

Fig. 3.1. Hugoniots of metals of the first group: (1) Mo; (2) Ir $(D+1)$; (3) Au $(D+1)$; (4) W; (5) Re $(D-2)$. Data by foreign researchers (dots); data from Russian centers (open circles).

transitions have been studied in other metals discussed below.

The second group (Fig. 3.2) includes fourteen metals whose Hugoniots are parabolic:

$$D = s_0 + a_1 U + a_2 U^2; \qquad a_1 > 0, \qquad a_2 < 0.$$

These are Al, Cr, Ni, Cd, Zn, Ag, Cu, Pd, Pt, Lu, Pb, Cs, Ce, and Te. The initial slopes of their Hugoniots are very different from the high-pressure limit (the minimum slope of 1.55 is that of Cr and Pt, the maximum slope of 1.9 was detected in Ce). The Hugoniot slope, D'_U, of most metals of this group changes in the experimentally studied pressure range. These are, for example, Al, Cu, Cd, Pb, and some other metals.

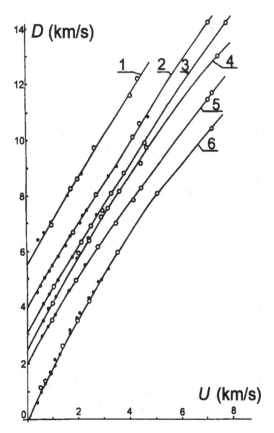

Fig. 3.2. Hugoniots of metals of the second group: (1) Ni $(D + 1)$;
(2) Cu; (3) Zn; (4) Cd; (5) Pb; (6) Ce $(D - 1)$. Experimental points are
marked by the same symbols as in Fig. 3.1.

The third group (Fig. 3.3) incorporates the metals whose
Hugoniots are also parabolic, but with positive coefficients s_0, a_1,
and a_2. These are such metals as Na, V, Nb, Ta, Co, and Rh.
The initial slope of their Hugoniots, D'_U, is 1.1–1.3, therefore one
should not expect a considerable change in this parameter due to
the transition to the high-pressure limit.

Many of the Hugoniots of the second and third groups (Al, Ta,
Zn, V, Cd, Ag, Cr, Cu, Ni, etc.) can be approximated by straight
lines in some ranges of kinematic parameters D and U, starting
from s_0 $(U = 0)$. Copper, aluminum, and nickel have the widest
linear ranges (from s_0 to 11, 12, and 11.0 km/s, respectively). For
other elements the range of linearity is quite narrow, and for some

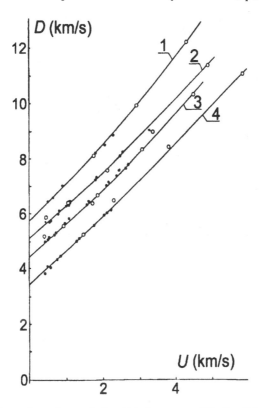

Fig. 3.3. Hugoniot plots of the third group: (1) Co $(D + 1)$; (2) V; (3) Nb; (4) Ta.

of them the Hugoniots are linear only in a very narrow range of $s_0 < D < D_b$.

The fourth group (Fig. 3.4) includes seventeen elements whose Hugoniots can be approximated by two crossing straight lines with different slopes ('broken' Hugoniots). These are rare-earth elements of the sixth period of the periodic system (La, Pr, Nd, Sm, Gd, Tb, Dy, Ho, Er, Tu, Yb, and Lu) and also Y, Rb, Sr, Ca, and Sc. The slopes of the lower sections of all the Hugoniots, except that of Ca, are less than unity, the slopes of the higher sections range between ≈ 1.0 (Y, Sm, Dy, Ho, and Tu) and 1.7–1.8 (La and Pr). The distinctive feature of calcium is the upwards curvature of its $D - U$ curve, i.e., the slope of the lower section is larger than that of the higher section. In the studied pressure range, the slopes of the second sections are constant (they are nearly rectilinear). Apparently, the slope should change at higher

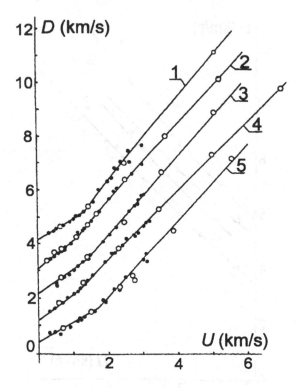

Fig. 3.4. Hugoniots of the fourth group: (1) Pr $(D+2)$; (2) La $(D+1)$; (3) Nd; (4) Gd $(D-1)$; (5) Y $(D-3)$. Data by foreign researchers (dots); data from Russian centers (open circles).

shock parameters. The small slope of the lower Hugoniot portion is characteristic of high-compressibility states, and beyond the cusps the Hugoniots switch to higher slopes corresponding to a lower compressibility (higher D_U'). These shapes of Hugoniots derive from the electronic structures of these metals [33], in which internal d-shells are not filled and external s-electrons can be transferred by compression to internal d-shells that results in a high compressibility of s-shells. Because of these transitions of s-electrons to d-shells, atomic configurations with closed internal shells, characterized by a lower compressibility, are formed. The switch-over from the flatter Hugoniot section to the steeper one corresponds to the completion of electronic transitions. Since, in their initial states, most of the metals of this group have structures in which atoms are packed densely, the sharp decrease in their compressibilities on the higher Hugoniot sections cannot be accounted for in terms of transition to a denser phase. At the same time, we should note that the presentation of experimental data in the form of two straight lines is only an approximation due to finite measurement uncertainties and densities of experimental points. It seems that transitions between electron shells should result in smoother curves between two linear portions of Hugoniots.

The last, fifth group includes those elements in which compression causes transitions from initially loose to denser states. The most extensively studied transition of this kind is the $\alpha \to \varepsilon$ transition in iron. This group also includes Bi, Ti, Sn, Zr, Hf, Eu, and Pu. The Hugoniots of these materials (Fig. 3.5) consist of three sections. The first one describes the compression of the initial phase, the second horizontal portion corresponds to the phase

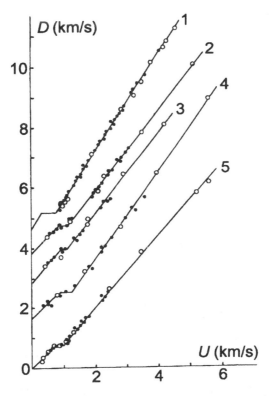

Fig. 3.5. Hugoniots of the fifth group: (1) Fe; (2) Zr; (3) Hf; (4) Eu; (5) Ti $(D - 5)$.

transition, i.e., the range in which the initial and denser phases co-exist, and the third section corresponds to a Hugoniot of a denser metal phase.

This classification of Hugoniots is somewhat modified in comparison to that given in Ref. [32]. The classification is not quite exact because experimental data about some metals are scarce and insufficient for a certain classification of their compression curves. From this viewpoint, it is desirable to obtain more accurate data about Ba and Ga Hugoniots, which are preliminarily classified with the first group, and about the lower portions of the Hugoniots of Sc and some metals of the second, third, and fourth groups.

Now let us discuss relative measurements of the compression of metals from various groups of the periodic system. First of all, let us compare Hugoniots of metals of the three large periods. They are plotted in pressure–compression coordinates in Fig. 3.6. The most striking difference is observed between the compressibilities of alkali and transition metals. For example, the compression of potassium at $P = 100$ GPa is $\sigma = \rho/\rho_0 \simeq 3.5$, whereas that of nickel, transition metal of the same period, is $\sigma = 1.3$, i.e., a factor

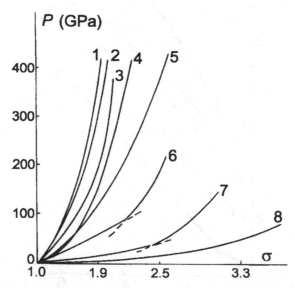

Fig. 3.6. Relative compressibilities of metals of period IV in the periodic system: (1) Ni; (2) Cr; (3) V; (4) Zn; (5) Ti; (6) Sc; (7) Ca; (8) K.

of 2.7 smaller. The compression of cesium at $P = 50$ GPa is a factor of 2.9 higher than that of iridium, the transition metal of the same period (Fig. 3.7). A natural explanation of this is the

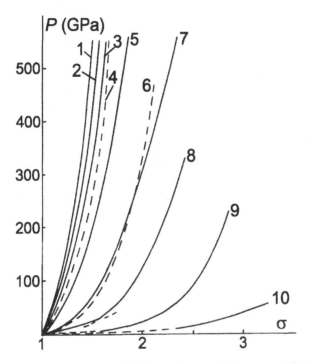

Fig. 3.7. Relative compressibilities of metals from period VI of the periodic system: (1) Ir; (2) Re; (3) W; (4) Au; (5) Ta; (6) Tl; (7) Pb; (8) La; (9) Ba; (10) Cs.

difference between electronic structures of these metals. On one side are alkali metals with easily compressible s-shells containing one electron loosely bound to a nucleus, on the other side are densely packed transition metals with filled electron shells and, hence, with a considerably smaller compressibility.

3.2   Absolute measurements of metal compressibility in underground nuclear tests

Absolute measurements were performed under much higher pressures during underground nuclear tests. The most advanced results in this field are measurements of iron compression at a pressure of up to 10 TPa [15] and of aluminum at 1.7 TPa [21].

Underground nuclear explosions may generate considerably higher pressures, which has been demonstrated many times and used in relative measurements [18, 20], but measurement uncertainties of shock parameters and errors of various factors introduced to take into account, in particular, sample preshock heating under these conditions are so high that uncertainties of experimental points may be comparable to discrepancies between calculations using different theoretical models.* In this sense, the highest pressures applied to iron and aluminum quoted above are close to the compression limit, and different theoretical models yield different predictions about whether the compressibility will change at a higher pressure. This matter will be discussed in detail below. Here we will consider measurements of the compressibility of iron and aluminum, the two metals used as standard materials in dynamic experiments (drivers in the impedance-matching technique).

3.2.1 Shock compression of iron in a pressure range of 4–10 TPa

The discussed measurements of iron compression were performed in 1970–74. We used the deceleration technique, i.e., a steel disk (impactor) was accelerated to a velocity of several tens of kilometers per second and generated the required compression in a steel target on impact with it. The experimental setup is shown in Fig. 3.8 [11, 14, 15]. Two versions of the accelerator were used in our experiments. In both cases, the impactor–target system was identical: a steel disk 25 mm thick and 700 mm in diameter hit a target 50 mm thick. The target was covered with a 5 mm Fe-shield on the side of the approaching impactor. The function of the 250 mm styrofoam spacer was to moderate the rate of pressure acting on the impactor plate and to generate a small additional pressure due to the evaporated polystyrene. The velocities of the impactor and shock in the target were measured using electrical shorting-pin detectors not susceptible to the relatively strong shock generated in air ahead of the flying impactor. The detectors were placed symmetrically about the axis of the impactor and target, two to four sensors being introduced to each cross section. The number

---

* Specifically, portions of Hugoniots in the regions close to the compression limit.

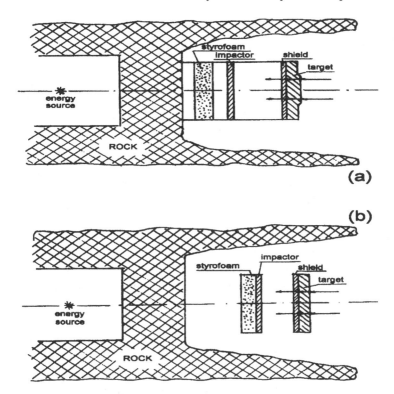

Fig. 3.8. Two configurations of full-scale experiments in which the impactor and shock velocities are measured.

of cross sections was two to four on the impactor's path and three to five in the target.

Before discussing the measurements, let us consider the demanded parameters of the accelerating system.

1. The impactor velocity at the moment of its impact with the target should be close to double the particle velocity in the target, i.e., $W = 2U$.

2. The impactor velocity should be constant when it approaches the target, i.e., in the collision regime $W = \text{const}$.

3. The impactor preshock heating due to waves generated in it and radiation from the nuclear explosion should be as low as possible.

4. The flying impactor plate should not disintegrate in flight and remain flat to a good accuracy.

The feasibility of the first three requirements was checked in a series of preliminary calculations of several versions of the device with system parameters varied over a wide range. These calculations took into account all the processes due to the nuclear explosion, such as generation of a shock in the rock, its propagation through the rock, spread of rock fragments and their delay in the polystyrene foam, shock propagation through polystyrene, impactor motion (its acceleration with due account of the additional thrust due to evaporated polystyrene), its impact against the target, and propagation of shock in it. The gas-dynamic stage of the shock propagation was calculated using the Mie–Grüneisen equations of state, polystyrene was described by the equation of state for an ideal gas with a polytropic exponent $\gamma = 5/3$ and $\rho_{00} = 0.03$ g/cm^3. It is, naturally, impossible to build a system in which the condition $W = 2U$ is fulfilled exactly, primarily because the impactor is preheated by the shock and its density is changed. For example, in the system shown in Fig. 3.8(a) and in one of two experiments with the configuration of Fig. 3.8(b), [11, 14, 15] the calculated parameters were as follows: the amplitude of the first shock in the impactor $P \leq 0.5$ TPa (the temperature $T \leq 2 \cdot 10^4$ K), the average density $\rho = 1.05\rho_0$ due to the additional thrust generated by evaporated polystyrene. Nothing can be done to change this situation because a thrust of some origin and intensity is always present in all cases. Its variations (for example, due to the polystyrene spacer position or its removal) lead to respective variations in the pressure in the impactor and to an additional acceleration of it, which is undesirable in many cases. Therefore, an optimal configuration was selected. The system was optimized by fitting W to $2U(\overline{D})$, where $\overline{D}$ is the average shock velocity in the target. The velocities $\overline{D}$ and W were determined empirically, and the function $U(\overline{D})$ was derived from the iron equation of state used in the calculation.

Figure 3.9 shows as an example the calculated velocity W versus the coordinate for the configuration of Fig. 3.8(a). You can see that the plate is accelerated in bursts when shock waves reflected from its back surface reach its free surface. But in the process the amplitude of the jumps scales down, and the impactor motion becomes more uniform. On the final portion of its path, the plate velocity becomes constant, and the projectile picks up about 90% of this value on the first quarter of its path.

With this configuration, we had the impact velocity $W_{\text{calc}} = 36.32$ km/s, the calculated average shock velocity in the target

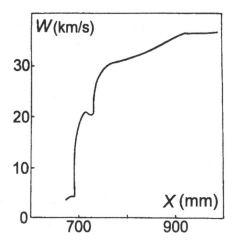

Fig. 3.9. Calculated velocity of the impactor front surface versus coordinate.

$\overline{D} = 29.1$ km/s, and, respectively, $U(\overline{D}) = 18.1$ km/s. Let us introduce a parameter to be optimized, $\alpha = U(\overline{D})/W$. When W is exactly $2U$, $\alpha = 0.5$ and $U(\overline{D}) = W/2$. Under real conditions, we considered our system optimal when α deviated from $1/2$ within 0.5–1.0%. In this case, the experimental uncertainty was surely larger, and there was no need to introduce corrections that would degrade our confidence in the results.

In the case under discussion, $\alpha = 0.498$, i.e., the system was optimized, the correction due to the difference between W and $2U$ (on the shock front) was small and could be ignored, according to our condition of the system optimization.

In one of the experiments performed with the configuration of Fig. 3.8(b), the calculated system parameters were close to those of the configuration discussed above. In the other experiment at a considerably higher energy released from the source, the pressure in the iron target was a factor of two higher (10 TPa) than in the two previous experiments at the same acceleration path length.

These are some calculated parameters of this system: the rock pressure before the shock wave is launched into air is 2.5 TPa, the pressure generated by the first shock in the impactor is 0.4 TPa, which is even lower than in the configuration of Fig. 3.8(a) owing to a wider gap. Before the impact against the target, the impactor velocity is nearly constant. The optimized parameter $\alpha = 0.503$, i.e., fairly close to the optimal value.

The measurements of the shock kinematic parameters are given in Fig. 3.10, where the time of signal detection is plotted against the coordinate. On the left portion of the curve, measurements

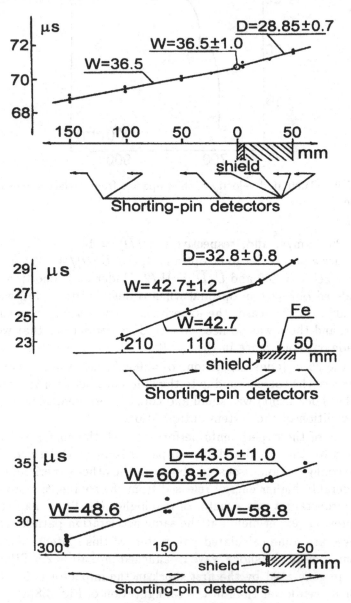

Fig. 3.10. Impactor and shock arrival times versus coordinate: experimental points (dots); impact velocity (open circles) on the target surface.

of the impactor motion are plotted, and the right portion illustrates the shock propagation in the target. In the first two experiments the velocity uncertainties are approximately equal, and in the third experiment the error is slightly larger. Note also that the impactor velocity, W, is well matched to the shock velocity on the air–target interface, $\overline{D}$, in the first two experiments. In the third experiment the impact velocity was derived by extrapolating the velocity $\overline{W}$ determined on the intervals between measurements taking into account the calculated impactor plate acceleration.

The experimental data are summarized in Table 3.1. They are compared to the laboratory measurements of the iron compress-

Table 3.1.

W_1 km/s	W_2 km/s	W_{av} km/s	D km/s	U km/s	P TPa	ρ g/cm^3
36.5	36.5	36.5 ± 1.0	28.9 ± 0.7	18.3	4.1	21.3
42.7	42.7	42.7 ± 1.2	32.4 ± 0.8	21.4	5.4	23.0
48.6	58.8	60.8 ± 2.5	43.5 ± 1.0	30.6	10.5	26.5

ibility and to relative measurements taken from Refs. [4, 7, 24, 27][†] shown in Fig. 3.11. There is good agreement among all results. The Hugoniot slope, D'_U, gradually changes with the shock velocity. At $D \simeq (8 - 10)$ km/s it is $D'_U \simeq 1.6$, and at $D > 15$ km/s $D'_U \simeq 1.20$. At higher velocities of up to $D \simeq 60$ km/s the slope is constant. Let us recall that $D'_U \simeq 1.20$ is approximately equal to the high-pressure limit, therefore one may expect that at higher kinematic parameters the slope will be approximately the same, i.e., at $D > 15$ km/s the $D - U$ curve of iron will belong to the first group in our classification of Hugoniots:

$$D = 6.41 + 1.213 \cdot U; \qquad \rho_0 = 7.85 \text{ g/cm}^3.$$

If the phase transitions in iron are ignored, which is admissible, given that the measurements were performed in the 100-Mbar range, all the absolute measurements of iron compression can be described by a curve composed of three sections with D and its derivative, D'_U, joined smoothly:

$$D_1 = 3.664 + 1.79 \cdot U - 0.03424 \cdot U^2, \qquad 1.4 < U < 8.0 \text{ km/s};$$

[†] Parameters derived from measurements by the authors of Refs. [4, 7] are listed.

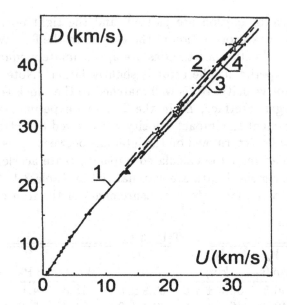

Fig. 3.11. Hugoniot plot for iron. Laboratory data (dots). Full-scale relative measurements: open circles, filled circles, triangles; full-scale absolute measurements from Refs. [14, 15] (crossed circles). (1) experimental data; (2) self-consistent field (SCF); (3) MHFS; (4) TFCK.

$$D_2 = 5.8695 + 1.239 \cdot U - 17.0 \cdot 10^{-5} \cdot U^2, \quad 8.0 < U < 22.0 \text{ km/s};$$
$$D_3 = 6.9885 + 1.190 \cdot U - 11.0 \cdot 10^{-5} \cdot U^2, \qquad U > 22.0 \text{ km/s}.$$

Figure 3.11 also shows Hugoniots calculated using the three most common theoretical models, namely, the Thomas–Fermi model (TF) for electrons with quantum and exchange corrections [34] taking into account the non-ideal properties of nuclei [35] (TF model corrected by Kopyshev, TFCK), the modified Hartree–Fock–Slater model (MHFS) [36], and the self-consistent field model (SCF) [37].

One can see that in the range of high shock velocity ($D > 20$ km/s) all three models yield parameters close to experimental data. In fact, they are equal to the measurements within the measurement uncertainty. Nonetheless, the TFCK model is preferable because its results are closer to experimental data at high pressure.

Now let us briefly discuss the theoretical models (Fig. 3.12). The Thomas–Fermi model with nonideal nuclei (TFCK) is statistical and yields monotonic curves of thermodynamic parameters plotted against ρ/ρ_0. Owing to its relative simplicity and

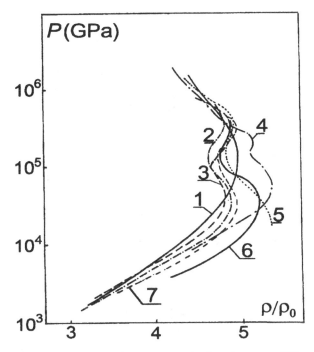

Fig. 3.12. Hugoniots of aluminum calculated using different models: (1) TFCK; (2) SCF; (3, 6, 7) versions of MHFS; (4) INFERNO; (5) ACTEX.

wide range of application (thermodynamic functions have been calculated for many chemical elements), it is widely used in calculations of asymptotic parameters at very high energy density, including shock compressibility of metals. We took the following initial conditions ahead of the shock front in our TFCK calculations of Hugoniots: ρ_0 was the initial (crystalline) metal density, $P_0 = 0$ was the initial pressure, and the initial energy, F_0, was taken equal to the empirical evaporation energy with the minus sign. These initial conditions are to some extent equivalent to the low-pressure asymptotic of a calculated curve fitted to experimental data, although they do not include direct measurements of compressibility. The TFCK model ignores the shell structure of atoms, but when atomic shells are consecutively ionized by a shock, this should result in smooth oscillations of Hugoniots and other thermodynamic functions against the background of statistical calculations, like those by the TF model. Such curves are derived from the Hartree–Fock model, which takes into account the exchange interaction in the Slater approximation [38], from the modified Hartree–Fock–Slater model (MHFS), from the self-consistent field (SCF) model, and others. The most developed models are the latter two. Calculations of thermodynamic functions by these models are cumbersome and labor-consuming, therefore only Hugoniots of metals used as standards, namely aluminum, iron and lead, have been calculated.

The MHFS model allows some approximations leading to the Hartree–Fock–Slater model [38] and hence to a considerable simplification of quantum mechanical calculations. The basic two simplifications are:

- the description of electrons in the mean atom approximation;

- a quasi-classical approach to the contribution of highly excited electrons that results in the so-called boundary of the quasi-classical spectrum, which allows one to vary parameters of calculated functions.

The feature of the SCF model, as compared to MHFS, is the quasi-continuum approximation, in which the upper and lower boundaries of energy bands are determined by equating to zero the wave functions and their derivatives, and the density of states in each band is calculated in the free-electron approximation.

Note that the development of both quantum-mechanical models –MHFS and SCF–is not complete, moreover, their progress is illustrated by different Hugoniots of aluminum with oscillations given in Fig. 3.12, whose amplitudes and peak positions calculated using different versions of MHFS do not coincide [36, 39]. By the way, the development of the TFCK model also cannot be considered as finished because higher-order terms in the interparticle interaction may change the Hugoniot position. Some corrections may also derive from the ion interaction, which sometimes cannot be easily incorporated in the scheme.

In this context, an important role is played by absolute measurements of shock compressibility at superhigh pressures, which can be directly compared to calculated Hugoniots and thus the best model can be selected, as described above, taking direct measurements of the iron compressibility.

For completeness, Fig. 3.12 also shows results of simulations using the ACTEX and INFERNO models developed in the USA, which take into account the shell structure of atoms. They are shown by the curves on the right of the diagram.

3.2.2 Compressibility of copper, lead, and cadmium under a pressure of up to 5.1 TPa and molybdenum up to 1.5 TPa

Compressibilities of these metals were measured by the impedance-matching technique, the iron–lead system being the first in which the psychological 1-TPa barrier was overcome in 1966 [4]. At that time, these and subsequent measurements of compressibilities of Pb–Cu–Cd [6, 7], Fe–Pb [7], and Fe–Mo [11, 40] systems were relative, and only after subsequent determination of the iron Hugoniot by the impact technique [14, 15] could these data be expressed in absolute terms.

The experimental setup is shown in Fig. 2.6. With this configuration, it was sufficient to have the equation of state (or, at least, the Hugoniot) of the first material in order to determine compressibility parameters of all metals placed downstream of the standard material when the layers of tested metals were packed one on top of another.

In these experiments, we used electric shorting-pin detectors like those in iron compressibility measurements. There were four to six sensors on each level (interfaces between metal disks). The signals were recorded by fast oscilloscopes with a resolution of $\pm 5 \cdot 10^{-9}$ s.

From the data obtained in two experiments on the Fe–Pb system, we derived shock velocities, and with small corrections for effects due to interfaces they are

I $D_{\text{Fe}} = 25.70$ km/s; $D_{\text{Pb}} = 20.72$ km/s;

II $D_{\text{Fe}} = 31.90$ km/s; $D_{\text{Pb}} = 26.12$ km/s.

These are direct measurements to be compared to laboratory data, which is done in Fig. 3.13. You can see that both series of data are quite compatible, which indicates their correctness. Note that this procedure of checking the correctness of data was used by many researchers [4, 5, 18, 41] because it is fairly simple and reliable. The experimental $D - D$ curves are almost linear ($D_{\text{Fe}} = 0.400 + 1.343 \cdot D_{\text{Pb}} - 3.49 \cdot 10^{-3} \cdot D_{\text{Pb}}^2$; $4 \leq D_{\text{Fe}} \leq 60$ km/s, $3 \leq D_{\text{Pb}} \leq 53$ km/s), and the shock compressibility of the tested material can be expressed in terms of the shock velocity in the standard material throughout the studied range of shock velocities. Through these relationships, new statistical data can be derived for further analysis. The $D - D$ variables can be changed for thermodynamic compressibility parameters using $P - U$ dia-

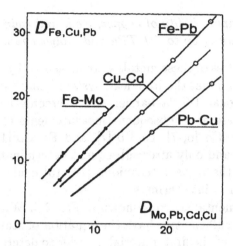

Fig. 3.13. Comparison of shock velocities on the shield-tested metal interface $(D - D$ diagram): laboratory data (dots); full-scale measurements (open circles).

grams in accordance with the routine of the impedance-matching technique, which uses the equation of state of the standard material.

The basic compressibility parameters of lead derived from $D_{Fe} - D_{Pb}$ measurements are listed in Table 3.2. The starting parameters of the shield (iron) were derived from measurements of D_{Fe} and the parabolic $D - U$ curve given in the previous section. The two points for lead in the terapascal region combined with the laboratory data define its $D-U$ curve, which is described in the high-pressure range $(14\,\mathrm{km/s} < D < 30\,\mathrm{km/s})$ by the linear function

$$D_{Pb} = 3.19 + 1.167 \cdot U; \qquad \rho_0 = 11.34 \ \mathrm{g/cm^3}.$$

Given the Hugoniot of lead, we can now analyze the data of two experiments with the Pb–Cu–Cd system, in which lead, in accordance with the impedance-matching routine, is the driver for the copper sample placed downstream of it. The system is triple, therefore the Hugoniot of copper, which is a driver for cadmium, is needed to determine the cadmium compressibility.

After the required calculations and transformations using the $P - U$ diagram, two points were plotted on the copper Hugoniot (Table 3.2). These data are plotted in $D - D$ coordinates in Fig. 3.13. As with the previous Fe–Cu pair, the interpolation

Table 3.2.

Parameters in standard material		Tested material	Compression parameters in tested material			
D km/s	U km/s	ρ_0 g/cm^3	D km/s	U km/s	P TPa	ρ g/cm^3
Fe, 25.70	15.91	Pb, 11.34	20.72	15.02	3.53	41.22
Fe, 31.90	21.01	Pb, 11.34	26.12	19.76	5.85	46.57
Pb, 14.65	9.815	Cu, 8.93	17.82	9.96	1.59	20.25
Pb, 22.17	16.26	Cu, 8.93	26.18	16.67	3.90	24.58
Cu, 17.47	9.725	Cd, 8.64	16.08	10.14	1.41	23.40
Cu, 25.30	15.90	Cd, 8.64	23.88	16.41	3.39	27.62
Fe, 17.25	9.20	Mo, 10.07	16.10	8.58	1.39	22.06

between the two pressure ranges in the $D - D$ diagram is practically linear. In the range of 15 km/s $< D <$ 27 km/s, the copper Hugoniot is described by the linear function

$$D_{Cu} = 5.137 + 1.268 \cdot U; \qquad \rho_0 = 8.93 \text{ g/cm}^3.$$

The parameters of the cadmium compressibility (Table 3.2) were derived from this function and $D_{Cu} - D_{Cd}$ measurements. These data are also plotted in Fig. 3.13. They can be similarly linked to the laboratory measurements.

As an example, Fig. 3.14 shows measurements of the shock velocity in the Pb–Cu–Cd system performed in 1970 in $D - X$ coordinates. The instantaneous velocities on the interfaces were

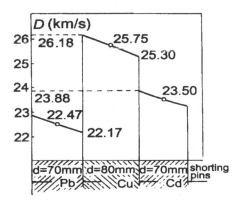

Fig. 3.14. Calculation of shock velocities on interfaces from experimental data.

calculated using average shock velocities.

The experimental point on the molybdenum Hugoniot was measured in the system where the driver was iron. The parameters
of the molybdenum compressibility (Fig. 3.13 and Table 3.2) were
derived from measurements using the routine of the impedance-
matching method. It is noteworthy that the difference between the
slopes of the lines drawn through the points of the full-scale experiment and laboratory points, D'_U, is larger than in previous cases.
The cause of this is not yet clear, and it is advisable to obtain new
measurements at pressures higher than in previous experiments.
Studies of such states are of a certain interest because the $D - U$
curve of molybdenum is linear, and its slope, $D'_U = 1.2$, coincides
with the theoretical value calculated by the TFCK model in the
high-pressure limit. Therefore, we may assume that the $D - U$
curve of molybdenum is rectilinear throughout the entire range of
shock velocities, going from the sound velocity and up to hundreds
of kilometers per second.

Until recently measurements at extremely high compression
parameters ($P > 1$ TPa) could be performed only in full-scale
experiments, which are impossible at present because of the total
ban on nuclear explosions. There has been some progress, however, in this pressure range [31] since we built a new generator of
shock waves which produces extremely high compression of tested
materials. The pressure generated in molybdenum by this device
($P \simeq 2$ TPa) is approximately equal to that in underground experiments [22]. Other parameters, specifically the shock velocity,
are different. The new measurements confirm the earlier assumption [32] that the $D - U$ curve of molybdenum is linear. Compression parameters of some metals, including Mo, measured at
maximum shock parameters produced in the laboratory are listed
in Table 3.3.

Figure 3.15 shows $D - U$ Hugoniots of metals measured both
in laboratory and in underground nuclear tests. In addition to
full-scale experiments described above, this plot includes the data
from Refs. [9, 14–16, 18–27, 42, 43]. Laboratory data were taken
from Refs. [28, 44–47].

Starting with the shock velocity corresponding to the highest
pressure available in the laboratory, the Hugoniot slopes of most
metals change from $D'_U \simeq 1.5$–1.7 to $D'_U \simeq 1.20$, and $D - U$
curves become approximately parallel. Tungsten and molybdenum, whose Hugoniots are first type, have curves rectilinear

Table 3.3. Impedance-matching experiments. Shock parameters in the shield (iron): $D = 20.19$ km/s, $U = 11.54$ km/s, $P = 1.83$ TPa

Studied metal	ρ_0 g/cm^3	Shock parameters			
		D km/s	U km/s	P TPa	ρ g/cm^3
Molybdenum	10.2	18.74	10.74	2.05	23.90
Aluminum	2.71	24.17	15.08	0.99	7.21
Tantalum	16.38	15.85	9.36	2.43	40.00
Titanium	4.5	20.95	13.67	1.29	12.95

throughout the studied range of wave velocities. We obtained the following slopes of metal Hugoniots: 1.19 (Fe), 1.15 (Pb), 1.20 (Ti), and 1.21 (Cu). Let us add the data from Fig. 3.15: 1.26 (Mo) and 1.19 (Al). Thus the average slope is about 1.20. The equal slopes of the curves of all metals indicate that in this range they lose their individual properties. This is demonstrated most clearly by the diagram of Fig. 3.16, where atomic volumes V are plotted against the atomic number Z at several pressures. At a low pressure the curves are periodic with the atomic number because

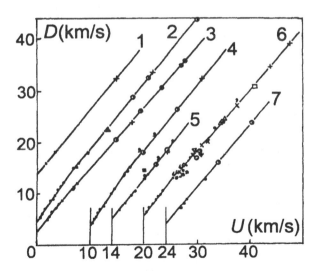

Fig. 3.15. Hugoniots of metals: (1) W $(D+10)$; (2) Fe; (3) Pb; (4) Cu; (5) Mo; (6) Al; (7) Cd. Laboratory data: dots, stars, upcommas, open circles; measurements of full-scale experiments: half-filled circles, pluses, crosses, triangles, squares, downcommas, filled circles.

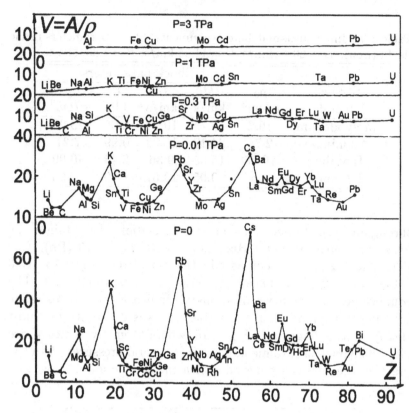

Fig. 3.16. Atomic volumes of chemical elements at several pressures.

electrons consecutively fill atomic shells. Alkali and alkaline-earth metals, whose points are at maxima of the $V(Z)$ curve, have larger atomic volumes because they have weakly bound electrons in s-shells. These are metals with a looser electronic structure. The elements on the lower portions of the curve are transitional metals, which have many electrons in d-shells.

At a higher pressure, the differences between atomic volumes become smaller, the curve $V(Z)$ is more smooth, and in the high-pressure limit it is a straight line. At a pressure of 1 TPa its oscillations are weak, and at a higher pressure of 3–5 TPa the function $V(Z)$ is practically linear. One may assert that under these conditions chemical elements lose their individual properties since differences between energy spectra of solids are eliminated. Figure 3.16 shows that alkali and alkaline-earth metals, corresponding to maxima on the atomic volume diagram, have the highest compressibilities, and transition metals and those with densely packed lattices have the lowest compressibilities. On the curve of atomic volumes, these elements correspond to lower values. Within each period, the compressibility of both ordinary and transition metals increases with the atomic volume.

Metals of different periods with similar structures and belonging to lower Z portions of the periods (in particular, alkali and alkaline-earth metals) have higher compressibilities at higher atomic numbers.

No clear tendencies for compressibility curves of metals with smaller atomic volumes have been observed. This is apparently caused by the complicated character of electron interactions under shock load, in particular, due to differences among structures of these metals at low pressure.

3.2.3   Compressibility of aluminum in a range of 0.2–1.7 TPa

Besides iron, aluminum is another standard material used as a driver in impedance-matching experiments. Whereas iron shields are used in experiments with materials whose densities $\rho_0 >$ 5 g/cm^3, aluminum is employed at lower densities. The number of natural materials with $\rho_0 <$ 5 g/cm^3 is much larger than that of denser materials, therefore aluminum is a more important metal in this sense.

Aluminum compressibility has been studied in the laboratory by many researchers [9, 28, 30–33, 46, 47] under pressures of up to 1.0 TPa (Table 3.3). This value is close to the limit accessible in modern laboratory shock facilities.

In full-scale experiments, the pressure range of 0.4 to 1.7 TPa has been studied by absolute measurement techniques [9, 18–21] (the compressibility was also estimated at 3.0 TPa [20]) and compression of the Fe–Al system was measured by relative methods in the range of up to 400 TPa [18]. The shock pressure experiments used both the γ-reference method and the reflection technique.

The first γ-reference measurements were, naturally, performed by the authors of the method [19, 20]. Measurements performed by our group were reported in Ref. [21]. In order to implement the technique, a lot of technical problems, which were described in detail in Ref. [20], had to be solved. Let us discuss the basic points of the method.

Firstly, there was the selection of the γ-active reference material. For experiments with aluminum, europium was chosen [‡] because its nuclei have a neutron capture cross section a factor of about one thousand larger than that of aluminum nuclei. At this ratio of cross sections, the γ-quanta generated in the source could be detected against the existing radiation background.

Another topic was the correct account of sample preheat due to the flow of neutrons from the nuclear charge and γ-quanta from the reference samples. In the experiments reported in Ref. [20], the preheat led to a decrease in the aluminum density of 5 to 10%, which should have been corrected for. The optimal neutron energy for the (n, γ) reaction in europium is $E \leq 100$ eV, therefore the shock launched into the sample had to be delayed in order to bring down the neutron energy to the required level since most of the neutrons of the primary beam have higher energies.

Finally, it was necessary to investigate the shock front distortions due to pellets of the reference material inside the sample; the equality between the reference velocity and particle velocity in Al; the conditions under which the shock front is plane (symmetrical) and is delayed for a required time with respect to the primary neutron beam, etc.

The results of three γ-reference experiments with aluminum were reported in Ref. [20]. Two experiments were performed with similar configurations and, naturally, yielded close results (Fig. 3.15), the third one yielded only qualitative results since the uncertainties in kinematic parameters were too high ($\sim 10\%$).

Figure 3.17 shows the configuration of γ-reference experiments [21]. Its main difference from that reported in Ref. [20] is the larger separation between the nuclear charge and aluminum slab because of the higher explosion energy. Thus the effect of some parasitic factors on measurements was reduced and their interpretation was easier.

The aluminum slab was aligned strictly perpendicular with respect to the axis connecting the energy source and the slab center. A homogeneous cement spacer was placed upstream of the slab. Lead inserts were placed next to the aluminum block and collimator system in order to delay the shock wave in these directions. As compared to the configuration of Ref. [20], the pellet thick-

---

[‡] In fact, europium dioxide, whose density is close to that of aluminum, was used in experiments.

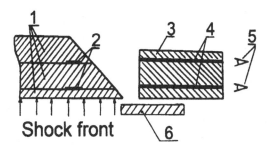

Fig. 3.17. Configuration of the γ-reference experiment: (1) specimen; (2) γ-sources; (3) collimator; (4) collimating slits; (5) γ-counters; (6) Pb shield.

ness was reduced by a factor of two to 2.5 mm. The pellets were placed outside the field of visibility through collimator slits but very close to it, therefore even after a small displacement of a pellet caused by the shock arrival, γ-quanta from the reference material passed through a collimator slit and were detected by photoelectric detectors, whose outputs were fed to oscilloscopes. With the configuration of Fig. 3.17, the average shock velocity was derived from the time of the shock transit between two reference pellets: the first had to be displaced to the second collimator slit so that its radiation could be also detected. We assumed that the reference material velocity was equal to the particle velocity in aluminum because the initial densities of both materials were equal and their Hugoniots at a low pressure ($P \leq 100$ GPa) were close to each other. The comparison between γ-reference and impedance-matching data indicated that this assumption was quite adequate.

The data of this experiment and of Ref. [21] are listed in Table 3.4. They are also plotted in Fig. 3.15, where they can be compared to laboratory data (the amount of experimental information about aluminum is so enormous that all available experimental

Table 3.4.

D km/s	U km/s	P TPa	$\sigma = \rho/\rho_0$	Ref.
24.2 ± 0.7	15.1 ± 0.4	0.99 ± 0.03	2.65 ± 0.10	[20]
23.4 ± 0.6	14.5 ± 0.3	0.93 ± 0.02	2.63 ± 0.07	[20]
40.0 ± 0.7	30.0 ± 2.0	3.20 ± 0.50	3.90 ± 1.20	[20]
30.5 ± 0.7	21.0 ± 0.6	1.73 ± 0.05	3.21 ± 0.13	[21]

points cannot be plotted in one graph without obscuring the general view) and results of full-scale experiments [18, 25].

All in all, there is fairly good agreement between laboratory and nuclear test data. Note that aluminum belongs to the few materials and is the only metal (excluding porous materials) whose laboratory and full-scale experiment data overlap.[§] This overlap takes place over a fairly wide interval of kinematic parameters and indicates the adequacy of experimental data obtained by both methods. If we limit ourselves to the interval of 11 km/s $< D <$ 80 km/s, the Hugoniot can be plotted in $D - U$ coordinates as a straight line:

$$D = 5.90 + 1.19 \cdot U; \qquad \rho_0 = 2.71 \text{ g/cm}^3.$$

In the high-pressure region, the aluminum Hugoniot was drawn through the points based on both experimental data and Thomas–Fermi calculations, which are in good agreement with the experiment. Throughout the entire range of shock velocities, the data can be approximated by a second-type Hugoniot

$$D_{\text{Al}} = 5.20 + 1.239 \cdot U - 0.7 \cdot 10^{-3} \cdot U^2.$$

To sum up, we have discussed investigations of the shock compressibility of metals, including the two main standard materials–iron and aluminum. There is a common tendency for all metals, namely, in the high-pressure limit their Hugoniots have a slope $D'_U \simeq 1.2$–1.3. In other words, starting from some shock velocity, their Hugoniots are approximately parallel straight lines.

In the following section we shall consider relative measurements of compression under superhigh pressures in underground nuclear tests.

3.3   Comparison of laboratory data on metal compressibility to measurements performed in underground nuclear tests

First of all, note (Fig. 3.15) that the coincidence between metal parameters measured in both laboratory and underground experiments is quite good. To be exact, a short-range extrapolation of laboratory $D - U$ curves to the region of full-scale experiments

---

[§] See the next section for details.

is usually in satisfactory agreement with full-scale measurements. The only exceptions are molybdenum [42, 43] and copper [43]. The deviation of data from Ref. [43] is, apparently, due to the aluminum shield used in the experiment, therefore the parameters of states of heavy metals were calculated using double compression Hugoniots, which had not been calculated to sufficient accuracy. Besides, the aluminum Hugoniot described by the equation given in Ref. [43] also deviates from experimental data. Poor measurement accuracy is indicated in Ref. [42]. Otherwise the full-scale and extrapolated laboratory data are in good agreement. Note, however, two essential differences between shock waves generated in laboratory and underground nuclear tests. Firstly, in the laboratory a shock with a pressure of up to 1–2 Mbar constant with time in some interval is generated, and in underground explosions a nonstationary divergent shock wave is generated.

The second difference is the time scale of experiments. Whereas in laboratory facilities the duration of shock action on a sample is usually a fraction of a microsecond, in underground explosions it is up to tens of microseconds or more. This difference is, naturally, due to different thicknesses of tested samples at equal velocities of shock waves, namely, 3–6 mm in the first case and 60–100 mm in the second case. The difference between these two experiments becomes quite notable under a pressure higher than 2–3 Mbar. Under these conditions, laboratory facilities usually generate converging shock waves, i.e., the pressure rate has the opposite sign, and the pressure rises faster as the wave travels from its origin and converges at the center, whereas a divergent wave becomes 'more plane' with distance and has a lower pressure rate.

We cannot exclude the possibility that the coincidence between extrapolated laboratory and full-scale measurements is accidental because shock parameters and sample dimensions in both types of experiments are different. Therefore the ranges of laboratory and underground experiments should, first of all, overlap. If the time dependence is correctly included in the scheme and the kinematic and thermodynamic parameters of compression are equal, the results in both cases should be identical, at least when the material does not undergo any chemical or structural transformations.

These topics always attract researchers' attention and are given a lot of consideration. In particular, in laboratory conditions, measurements with devices generating plane shock waves are performed in the same ranges of parameters as in experiments with converging waves. Experiments with underground explosions (testing-ground experiments) [11, 48] had similar aims. But results of these experiments were not fully correlated to other measurements, and we will try to fill this gap in this and the following sections.

In correlating the compressibilities of metals measured in full-scale and laboratory experiments, the researchers had several tasks. The first-priority problem was to verify some experimental points obtained in both laboratory and testing-ground experiments by comparing the measurements. There were two more urgent topics.

The first task was to verify the data on aluminum compression obtained by the γ-reference technique in underground explosions [19, 20, 21]. Let us recall that a reference pellet is presumed to travel at a velocity equal to the aluminum particle velocity. Dedicated experiments were undertaken to check whether this assumption was true. The problem was the more acute given that the experimental points under discussion determined the aluminum Hugoniot in the pressure range of 0.4 to 1.0 TPa, in which the experimental data raised most serious doubts with the researchers. Another issue was to verify the molybdenum Hugoniot curves at pressures of 1.4 to 2.0 TPa. Under these pressures, the data reported by different American researchers were highly controversial [22, 43], and the respective points were plotted on different sides of the linear extrapolation of the molybdenum laboratory curve [32]. Besides, it was desirable to check that measurements were independent of the scale of an experiment in studies of porous metals (Section 3.5).

The results of these comparisons were as follows. The identity between laboratory and testing-ground measurements is shown in Figs. 3.15 and 3.18, in which aluminum is taken as an example [9, 11]. On the middle portion of the $D - U$ curve at a shock velocity D ranging between 13 and 19 km/s, underground measurements were performed by the conventional technique, when tested aluminum samples were placed downstream of a metal (steel) shield set in the rock normally to the direction of the shock propagation. All in all, the agreement between the laboratory and testing-ground data is fairly good, although some experimental points deviate from the curve of averaged data too far–beyond the admissible uncertainty of 1–1.5%. The cause of this deviation was discussed in Refs. [9, 11]. In order to check the points on the

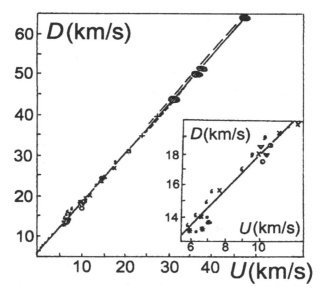

Fig. 3.18. Hugoniot of aluminum. Laboratory measurements: dots, upcommas, stars, downtriangles, open circles; full-scale measurements: crosses, downcommas, uptriangles, squares, pluses. Ellipses are areas of parameters derived from data of Ref. [18]. Dashed lines present data from Ref. [49] (long dashes) and from TFCK calculations (short dashes).

aluminum Hugoniot derived from the γ-reference measurements [19, 20, 21] and to plot the molybdenum Hugoniot, we used a laboratory facility [31] generating shocks with record parameters in the shield, namely, $U = 11.54$ km/s and $P = 1.83$ TPa. In each of five experiments of this series, samples of tested metals (specifically, aluminum and molybdenum) and standard iron samples were placed downstream of an iron shield, and measurements were averaged. The particle velocity and pressure were determined, as usual, by solving the problem about the decay of a discontinuity on the shield–tested metal interface. The iron Hugoniot was derived from the data of Refs. [11, 30, 31] (see Section 3.2.1). The time dependence of the shock in the facility was taken into account by introducing small corrections to the calculation. The resulting data are compared to testing-ground measurements in Figs. 3.15 and 3.18 (see also Table 3.3).

Note that the experimental facility described in Ref. [31] could generate a pressure corresponding to the parameters of underground experiments in which these metals were tested. The experimental point for aluminum exactly coincided with one of the points recorded in testing-ground experiments [20]. Thus we came to the conclusion that both types of experiments yielded identical measurements, i.e., the assumption about the equality of the γ-reference velocity to the aluminum particle velocity behind the shock front was justified.

As concerns molybdenum, one can see in Fig. 3.15 that the new point lies on the linear extrapolation of the $D - U$ curve [32]. In combination with earlier data obtained at lower compressions [40], it defines the molybdenum $D - U$ curve in the studied range of kinematic parameters and confirms the assertion [32] that its Hugoniot is linear throughout the studied range of kinematic parameters. The data of Refs. [22, 43], however, are not sufficiently accurate to be used in our analysis.

Experiments performed at the 'terapascal' laboratory facility with aluminum and molybdenum taken as examples marked, in a sense, the end of the monopoly of nuclear experiments in the intermediate range between previous laboratory measurements and calculations based on high-pressure models.

3.4 Relative compressibilities of iron, copper, lead, and titanium

The range of absolute compressibility measurements of denser metals ($\rho_0 \geq 7$ g/cm^3) is limited to a pressure of about 10 TPa by technical parameters of measuring devices and conditions of nuclear tests [15]. As concerns relative measurements, in which only the velocity of a shock propagating across layers of materials is measured and the second kinematic parameter, namely the particle velocity, is not detected, these experiments can be performed at considerably higher kinematic parameters, given good accuracy of shock velocity measurements. This possibility was demonstrated many times in various experiments [16, 18, 50]. It is noteworthy that the shock pressure is measured fairly accurately in these experiments because $P = \rho_0 D^2 (\sigma - 1)/\sigma$, and at a compression σ close to its limit, the accuracy of pressure measurements is largely controlled by the uncertainty of the shock velocity.

In 1981 our laboratory measured the relative compressibility of Fe–(Pb, Cu, Ti) under pressures of up to 20 TPa [16]. Later, in 1984–86, the higher limit of pressure was further increased to the enormous value of 7.5 Gbar (750 TPa) [18]!

In this section we shall discuss in detail the experimental layout and relative compressibility measurements of four metals Fe–(Pb, Cu, Ti) performed in 1981 because in that case the shock pressure (20 TPa) was only a factor of two higher than in absolute measurements, and we hoped that the TFCK model of iron, which adequately described experimental data at $P \leq 10$ TPa, would not greatly contradict the measurements at double the pressure.

The experimental layout is shown in Fig. 3.19. The measuring device was in this case placed very close to the nuclear charge (energy source) and was insulated from it (in addition to the air gap) by a fairly thick layer of polyethylene with inserted lead sheets. These sheets acted as a neutron shield for samples and sensors and also to shape a shock wave with a time-independent profile.

In this experiment, the shield was iron, to be exact, a steel disk (of low-doped St3) with a diameter of 1000 mm and a thickness of 50 mm. The diameter of samples (cylindrical pellets) was about 250 mm.

The shock front was made symmetrical and parallel to the plane of the measuring device by carefully aligning it perpendicular to the axis connecting the source of energy to the center of the measuring device.

Since electric pins were not sufficiently reliable detectors of position owing to possible breakdowns of gaps between electrodes under $n - \gamma$ radiation, we used a novel measuring technique based

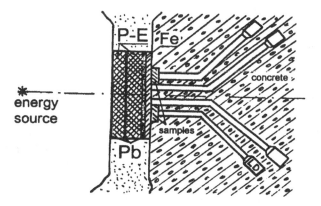

Fig. 3.19. Diagram of the full-scale experiment measuring relative compressibility in the Fe–(Cu, Pb, Ti) system under a pressure of 15–20 TPa.

on detecting light generated by the shock front on the interfaces between the shield and tested sample and between the sample and air (back side of the sample). Light was collected from sample surfaces not affected by relief waves propagating from the peripheral zone. Light was transmitted to detectors via steel pipes 30 mm in diameter. Their inside surfaces were polished to minimize the loss of light, and the inside volumes were evacuated down to a fraction of one torr so that air luminescence was not generated by neutrons and γ-quanta before light pulses due to shocks. Light detectors were protected from radiation generated by a nuclear charge by a thick layer of concrete. The distance to the light detectors was selected to ensure their reliable operation under realistic fluctuations of the radiation intensity. Light flux fed to a detector was converted to electric signals transmitted to fast oscilloscopes. According to the recorded data, the coincidence of light pulses generated on different portions of the shield–sample interface was fairly good (the jitter of pulses due to the shock recorded in four channels was about 10^{-8} s), i.e., the symmetry of the shock front hitting the shield was quite satisfactory. Another positive feature of the device from the viewpoint of data processing was that its structure was designed to minimize the attenuation of shock over the sample thickness, thus attenuation effects were ignored when interpreting experimental data.

Unfortunately, the uncertainty of determined shock velocities was rather large because of several factors, including the jitter of pulses on some oscilloscope traces. This uncertainty was 4% in Ti, 2.5–3.0% in Pb, 1.5% in Fe and Cu. The measured shock velocities in these materials were as follows: $D_{Fe} = 57.40 \pm 0.9$ km/s (standard material); $D_{Pb} = 48.79 \pm 1.2$ km/s; $D_{Ti} = 62.30 \pm 2.5$ km/s; $D_{Cu} = 55.87 \pm 0.8$ km/s.

The measurements are plotted in Fig. 3.20 in $D_{Fe} - D_x$ coordinates and compared to laboratory and full-scale experiments taken from Refs. [4, 7, 18]. As was stated above, the curves consist of two rectilinear sections joined by smooth curves in the intermediate range between the regions of laboratory and full-scale experiments, the slope of the low-pressure portion being higher than that of the high-pressure portion.

Data on the Fe–Pb system are more abundant. Figure 3.20 demonstrates that all the measurements are in fair agreement, the only exception is the data from Ref. [17], which are, apparently, affected by a serious systematic error. The data published by

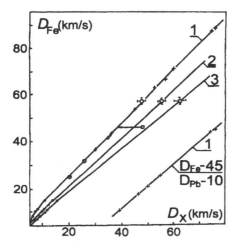

Fig. 3.20. $D - D$ curves of (1) Fe–Pb, (2) Fe–Cu, and (3) Fe–Ti. Absolute measurements: laboratory data (dots), full-scale experiments (open circles). Relative measurements: data from Ref. [16] (crossed circles), Ref. [17] (squares), and Ref. [25] (pluses).

Avrorin et al. [18] are really impressive because they cover a wide range of shock velocities and, judging by their coincidence with data from Refs. [4, 7] and by the shape of the $D_{Fe} - D_{Pb}$ curve, they adequately describe the state of this system under a superhigh pressure ($P > 60$ TPa).

In the range of shock velocities where all the three systems, namely, Fe–Ti, Fe–Cu, and Fe–Pb, were investigated, the $D - D$ ratios for these metals are the following:

$$Fe - Pb : \quad D_{Fe} = 2.35 + 1.125 \cdot D_{Pb}$$
$$Fe - Cu : \quad D_{Fe} = 0.86 + 1.01 \cdot D_{Cu}$$
$$Fe - Ti : \quad D_{Fe} = 1.60 + 0.89 \cdot D_{Ti}$$

(these relations are valid in the range 15 km/s $< D_{Fe} <$ 60 km/s).

In order to estimate the thermodynamic compression parameters of these metals, we shall use the iron Hugoniot calculated by the TFCK model. This is justified by the coincidence between calculations and measurements in the pressure range up to 10 TPa. But the more substantial evidence in favor of this model is that shock velocities calculated by this model using $D_{Fe} = 57.4$ km/s are very close to experimental data:

$$D_{Ti,calc} = 62.81 \text{ km/s} \quad D_{exp} = 62.30 \text{ km/s}$$

$$D_{\text{Cu,calc}} = 55.80 \text{ km/s} \qquad D_{exp} = 55.9 \text{ km/s}$$
$$D_{\text{Pb,calc}} = 49.59 \text{ km/s} \qquad D_{exp} = 48.8 \text{ km/s}$$

i.e., the maximum deviation in the case of lead is 1.5%, which is within the admissible experimental error. Thus the iron (shield) Hugoniot is calculated by the TFCK model. At $D_{\text{Fe}} = 57.4$ km/s the particle velocity $U_{\text{Fe}} = 42.3$ km/s and the pressure $P_{\text{Fe}} = 19.1$ TPa.

Since the shock parameters in this experiment were nearly constant with time, the compression parameters of Cu, Pb, and Ti can be derived directly from shock velocities. Given that the Hugoniots of lead and copper are close to that of iron and the titanium Hugoniot is sufficiently close so that the reflected Hugoniot technique can be applied to this case, the $P - U$ diagrams can be plotted using the mirror approximation. The resulting parameters are listed in Table 3.5 and plotted in Fig. 3.21. One can see that measurements of both laboratory and full-scale experiments are in fair agreement. The data from Ref. [27] for iron and from Ref. [26] for copper slightly deviate from calculated curves. The cause of deviations of the copper data was discussed above.

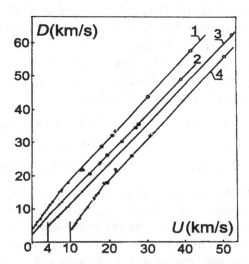

Fig. 3.21. Hugoniots of (1) iron, (2) lead, (3) titanium, and (4) copper. Absolute measurements: laboratory data (dots), full-scale experiments (half-filled circles). Relative measurements in full-scale experiments: data from Ref. [16] (open circles), Ref. [25] (pluses), Ref. [43] (downcommas), Ref. [27] (triangles), Ref. [17] (filled circles).

Table 3.5.			
Metal	D	U	P	ρ
	km/s	km/s	TPa	g/cm^3
Fe	57.4 ± 0.9	42.3	19.1	29.8
Pb	48.8 ± 1.5	39.3	21.7	58.0
Cu	55.9 ± 0.8	40.8	20.4	33.0
Ti	62.3 ± 2.5	48.6	13.6	20.5

To conclude this section, let us compare the parameters of metals measured in full-scale experiments to the curves calculated by the Thomas–Fermi model in the pressure range $P > 10$ TPa. These data are plotted in Figs. 3.22, 3.23, 3.24.

You can see that interpolated experimental curves for different metals approach Thomas–Fermi curves in a different manner. The Hugoniots of iron, lead, aluminum, and cadmium practically coincide with the calculated curves in the experimentally studied range, the Hugoniot of copper approaches the calculated curve from beneath, and that of molybdenum from above. But in all cases the experimental curves smoothly join the calculated ones, and not only the pressures, but their derivatives also approach high-compression calculations. This again indicates that the

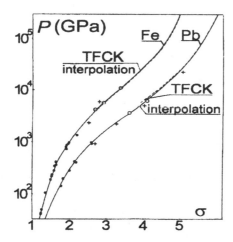

Fig. 3.22. Extrapolation of Fe and Pb Hugoniots to the region in which TFCK calculations are valid. Laboratory data (dots); full-scale absolute measurements (open circles); full-scale relative measurements (pluses).

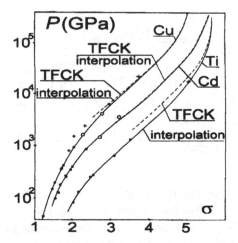

Fig. 3.23.   Same as in Fig. 3.22 for Cu, Cd, and Ti.

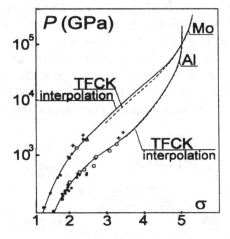

Fig. 3.24.   Same as in Fig. 3.22 for Mo and Al.

TFCK model yields parameters of metals in the high-compression limit.

In the next section we shall consider investigations of shock-compressed porous metals. Most experiments were performed in the laboratory, and measurements of Fe, Cu, and W parameters obtained in nuclear experiments are also given.

3.5   Compression of porous metals

The basic data used to construct equations of state are derived
from studies of shock compression of materials, primarily from the
compressibility of continuous bodies, whose density in the normal
state equals the crystalline density ρ_0. The derivation of their
thermodynamic parameters in considerably wider ranges of density and temperature than the neighborhood of a metal Hugoniot
curve originating from the point with the initial density ρ_0 demands, however, data about their behavior in a wider area of the
$P - T$ diagram. These data are obtained, in particular, by measuring the compressibility of porous metals, in which particles of
metal with the normal density, ρ_0, are separated by air or vacuum
gaps so that the average sample density is $\rho_{00} = \rho_0/m$, where m is
the degree of porosity. By varying the porosity, m, one can cover
with curves of shock compression a considerable area on the $P - \rho$
diagram between the Hugoniot curve of solid material ($m = 1$) and
a Hugoniot originating at a density $\rho \ll \rho_0$ ($m = 20 - 30$) [51].
The importance of studying the compression of porous materials
was first indicated by Ya.B. Zel'dovich [52]. There are many publications on properties of porous metals, the most notable of which
are Refs. [51, 53–56] because they contain more experimental data
than other papers and their interpretation in terms of equations
of state.

The impedance-matching technique is used in these studies.
The samples are cylindrical pellets with a diameter-to-thickness
ratio of more than three, compressed to a variable initial density
ρ_{00}. The dimension of separate grains is usually ≤ 100 μm. What
technical problems were resolved in developing the methods for
studying porous metals? The main topics are discussed below.

1. The effect of grain dimension.

The effect of the grain dimension on the propagating shock parameters is largely controlled by irregularities on the shock front.
It is admissible that at large dimensions of metal grains the shock
front be rough, and the question is, what is the maximum grain
dimension that does not affect the value of the shock velocity detected in experiment?¶

¶ That is the existing experimental techniques are not capable of detecting the
effect of shock front irregularities.

A direct answer to this question can be obtained by performing tests on one experimental facility at equal parameters of shock in the shield using samples of the same density, ρ_{00}, but with different grain dimensions. Usually experimenters test their facilities selecting grain dimensions at which the effect is presumed to be negligible (e.g., two experiments are performed with grain dimensions of 100 and 20 μm), and this assumption was confirmed by experiments [54].

An alternative way of addressing this issue is to conduct experiments with various sample thicknesses, but at a constant grain dimension. In this case, the effect should be larger in experiments with thin samples. The plate thickness in laboratory tests is varied by a factor of two to five. In full-scale experiments the sample thickness d was varied over a wider range. In the experiment performed by Trunin, Medvedev et al. [13] the variation in the thickness of porous copper samples ($m = 1.1, 3.2, 4.0$, $d = 80$ mm) was an order of magnitude larger than in laboratory experiments. We found that at grain dimensions of 100–200 μm or less the sample thickness did not affect measurements.

The third way is to record experimentally the profile of a shock in a porous sample and to compare it to a shock profile in a solid material at an equal pressure. It was found that the shock profiles in silicon dioxide powder with particle dimensions of 10–200 μm and in a piece of steel were identical [55] at $P \simeq 17$ GPa.

Note that in this discussion and in our considerations of other factors the statement about the absence of any effect depends on the measurement accuracy.

2. The effect of air on shock parameters.

The magnitude of this effect is determined in direct experiments performed at equal pressures, initial densities, ρ_{00}, and grain dimensions. The only difference is that in some experiments air is evacuated from samples to a pressure of a fraction of one torr, and in other experiments samples contain air at atmospheric pressure. The effect of air on measurements was not detected in experiments [55].

3. The effect of humidity on shock parameters.

Water vapor may condense on particle surfaces, which may be more important when particle dimensions are small and the total surface area is large, and have an 'additive' effect on the shock parameters. This effect was tested [54, 55] by comparing measurements of samples which were preliminarily baked and sealed, then placed in the measuring device, and of samples with identical parameters (ρ_{00}, grain dimension) not subjected to this treatment. The resulting measurements were identical.

These are major technical issues resolved in our studies of compression of porous metals and other porous materials (see below). To sum up, we proved that the effects of these factors on measurements of shock velocity were negligible. Therefore we considered the conditions of shock compression of porous materials to be close to equilibrium, so the same experimental techniques as in studies of solid materials could be used and the measurements could be easily interpreted.

In order to obtain reliable data at relatively low pressures ($P \leq 10\,\text{GPa}$), measurements were usually performed using piezoceramic transducers, which are more sensitive than commonly used electric shorting pins of copper wire with enamel insulation.

All in all, we measured the shock compressibility of sixteen porous metals. These were Mg, Mo, Cu, Fe, Cr, Ta, Pb, Ni, W, Al, U, Pu, Ti, V, Bi, and Zn. The studies of the compression of porous nickel (12 Hugoniots), copper (9) and molybdenum (8) were more detailed. The data about porous magnesium are scarce (only one porous Hugoniot). This section presents as examples the data about compressibility of nickel, copper, tungsten, and iron. These materials are selected because their properties were studied most comprehensively, in particular, the shock compression of nickel was studied at a record low initial density ρ_{00}, and porous Hugoniots of other metals were measured in both laboratory and full-scale experiments.

Figure 3.25 shows experimental curves for nickel. These are so-called 'equal-charge curves', i.e., shock velocities versus porosity recorded under equal parameters of shock in the shield of the device for each curve. The curves of other metals are similar, hence our conclusions concerning nickel are also applicable to other metals.

The most remarkable feature is that the curves of $D(m)$ are rectilinear beyond some value of m. Therefore we may linearly extrapolate the $D(m)$ curves beyond the porosity realized in experiments. The dropping slope dD/dm between the point $D(\rho_0)$, $m = 1$ and the rectilinear portion ($m \simeq 2$–3) is a function of the shock intensity in the shield, and the lower the intensity, the larger the slope.

The following two diagrams (Figs. 3.26, 3.27) display the ex-

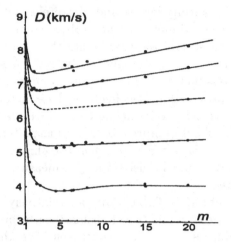

Fig. 3.25. $D - m$ diagram for porous nickel.

perimental data plotted in $D - U$ coordinates. The features of the curves plotted in these diagrams are more or less common to Hugoniots of other metals. The experimental points in Figs. 3.26, 3.27 are taken from Refs. [53, 54, 56]. One can see that the agreement

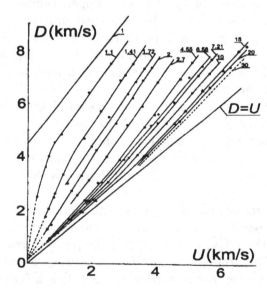

Fig. 3.26. $D - U$ diagrams of porous nickel (laboratory data). Numbers by the curves are respective porosities $(m = \rho_0/\rho_{00})$. Different symbols mark experimental points.

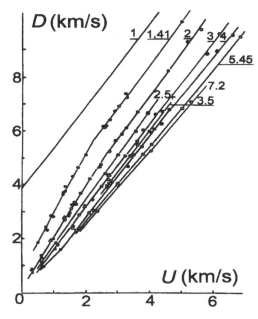

Fig. 3.27. $D - U$ diagrams of porous copper (laboratory data). Numbers by the curves are respective porosities $(m = \rho_0/\rho_{00})$. Pluses, filled circles, open circles, and triangles mark experimental points.

among all experimental data is fairly good, although some points deviate from curves plotted across the majority of experimental points.

Let us discuss in detail the technique used to measure the compressibility of porous copper, iron, and tungsten at a pressure beyond one terapascal in nuclear experiments. The shock generated by the underground explosion passed across an aluminum shield more than 100 mm thick and test samples with a thickness of 80 mm. We used powders of iron, copper, and tungsten with a content of the main component of 98.7%, 99.7%, and 99.8%, respectively, and grain dimensions of less than 100 μm. Since sample dimensions were comparatively large (the diameter was more than 250 mm), we could not vary the starting density over a wide range because it was difficult to fabricate samples of a uniform density. Therefore we experimented with samples of a density close to the so-called poured density, when a cell's volume was filled with powder and slightly pressed. Occasionally, the experimental points of the Hugoniot corresponding to this initial density and plotted on the $P - \rho$ diagram defined an approximately vertical line with a density $\rho \simeq \rho_0$. Given this set of states, the analysis of experimental results is quite easy because the pressures (energies) on a vertical Hugoniot centered at $\sigma = 1$ ($\rho = \rho_{0,\mathrm{cr}}$) are controlled by thermal components, of which the dominant component is electronic, and such an important parameter as the electron analogue of the Grüneisen coefficient, $\Gamma_e = 2/(m\sigma - 1)$, is controlled by the porosity m [53]. Since the initial densities of the powders (except tungsten) were close to the aluminum bulk density, the latter was selected as a shield material.

In these experiments the shock had a slowly dropping trailing edge, so when we substituted average values of the shock velocity with instantaneous velocities, the corrections on the aluminum-sample interface due to the shock decay were quite small (less than 1% of the average wave velocity $\overline{D}$).

As usual, the primary experimental data are plotted in Fig. 3.28 in $D - D$ coordinates. The plotted curves are rectilinear, those of iron and copper are close to one another, and the curve plotted through the tungsten points is steeper. The measurements at maximum pressures are listed in Table 3.6.

The data are plotted in $D - U$ coordinates in Fig. 3.29 and compared to laboratory and full-scale measurements of bulk samples. The diagram as a whole demonstrates good agreement among

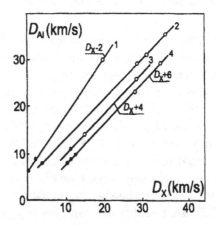

Fig. 3.28. $D - D$ curves for porous metals: (1) tungsten $m = 3.1$; (2) copper $m = 4.0$; (3) copper $m = 3.0$; and (4) iron $m = 3.4$ measured with aluminum shields. Laboratory measurements (filled circles); full-scale experiments (open circles).

Table 3.6.

Parameters in the shield		Studied material	Shock parameters in studied material			
D	U	ρ_0	D	U	P	ρ
km/s	km/s	g/cm^3	km/s	km/s	TPa	g/cm^3
29.90	20.17	W, 6.27	21.55	16.45	2.223	26.49
29.43	19.77	Fe, 2.27	28.27	21.28	1.365	9.18
23.20	14.54	Fe, 2.40	21.95	15.47	0.815	8.13
35.30	24.71	Cu, 2.23	34.83	26.52	2.060	9.34
31.21	21.27	Cu, 2.23	30.03	23.02	1.541	9.56
29.46	19.80	Cu, 2.23	27.67	21.60	1.332	10.18
26.10	16.97	Cu, 2.88	23.87	17.14	1.180	10.22

different measurements, which form a self-consistent pattern of Hugoniots of both porous and bulk metals in $D - U$ coordinates.

Figure 3.30 shows data about compression of porous copper at relatively low pressure (the range of laboratory experiments) generated by underground explosions. The aim of the experiment, as was noted above, was to study the effect of irregularities on the

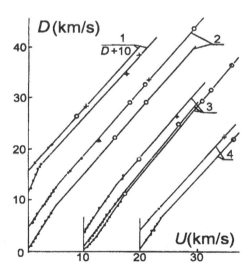

Fig. 3.29. Hugoniots of bulk and porous metals: (1) Mo $m = 1.0, 1.23$; (2) Fe $m = 1.0, 3.3$; (3) Cu $m = 1.0, 3.1, 4.0$; (4) W $m = 1.0, 3.1$. Laboratory data (dots), data of full-scale experiments: open circles, pluses, triangles.

Shock compression of metals

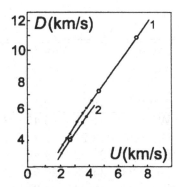

Fig. 3.30. $D - U$ diagram of porous copper. Laboratory data (dots); full-scale measurements (open circles).

shock front propagating across a highly porous material on the measured shock parameters. It is natural to suppose that the shock front 'roughness' controlled by dimensions of metal grains has the largest effect on measurements of small samples. Hence full-scale experiments with thicker samples should yield figures different from laboratory measurements. The coincidence between the two types of data would support the reliability of data about porous metals under terapascal pressures.

We tested porous copper samples with $m = 3.1, 3.2, 4.0$. Figure 3.30 indicates that both laboratory and nuclear-explosion measurements coincide. This leads us to the conclusion that within the ranges of grain dimensions used in our experiments and uncertainties of measured shock parameters, porous materials are compressed by shock in a quasi-stationary manner. The conclusion justifies full-scale experiments with porous metals under terapascal pressure.

We draw the following conclusions from the comprehensive analysis of experimental data:

1. Low-pressure sections of porous Hugoniots (Figs. 3.26, 3.27) form a fan of straight lines originating from the point $s_0' \geq 0$, $U = 0$ (or from a narrow interval $\Delta s_0'$). The value of the speed of sound s_0' varies between 0.15 and 0.30 km/s for the discussed materials. It is noteworthy that even at the low 10% porosity (Fig. 3.26, Ni curve) the Hugoniot with $m = 1.1$ originates at s_0'. The value of s_0 for air also falls in this interval.

2. Hugoniots with lower m have higher slopes, D_U', on their lower sections. Their slopes are significantly larger than that of

the Hugoniot with $m = 1$. The slope of a porous Hugoniot drops with m and at a sufficiently high porosity $(m \simeq 4)$ is equal to that of the bulk material. In the limit of high m the slope tends to $D'_U \simeq 1$.

3. At a given m, the slope D'_U is a function of the shock velocity. The dependence is stronger at smaller m. At high shock velocities the slopes of all Hugoniots become equal and tend to $D'_U \simeq 1.25$ (Fig. 3.29), i.e., equal to that of the bulk Hugoniot.

4. Hugoniots of the maximum porosity $(m > 5)$ have shapes such that extrapolation of their high-pressure portions to $U = 0$ yields negative s'_0, hence below some critical particle velocity, U_{cr}, such curves would yield a negative shock velocity. This result contradicts the conservation laws since in a shock wave the wave velocity is necessarily higher than the particle velocity. Therefore the slope of such a curve should decrease significantly in the range of low U and the curve should arrive at the common point s'_0. Nickel presents a typical example (Fig. 3.26) because detailed investigations of its properties at high m are available. The lower right line is described by the equation $D = U$ and is the boundary of all Hugoniots, even tangency with this line is forbidden by conservation laws. One can see in Fig. 3.26 that Hugoniots with $m = 15$ and 20 at $D \simeq 3.5$ km/s are very close to the line $D = U$, the distance between the curves at a constant U is only 200 m/s. This means that all the Hugoniots with $m > 20$ should be inside this narrow strip. Similar patterns can be plotted for other metals.

Now let us discuss experimental data plotted in $P - \rho$ coordinates (Figs. 3.31–3.33). You can see that measured Hugoniots densely cover the $P - \rho$ coordinate plane on the left of the $m = 1$ curve. The number of experimental points for nickel is the largest. Using also 'lines of equal charge' (Fig. 3.25), one can plot any Hugoniot in the studied range of m, including other metals.

At large m, a sample can be transformed by a shock to a state in which its density is not higher, but lower because of shock heating [51, 53, 54]. Such states were detected in some metals, including Ni, Cu, Mo, and others. The lowest density under shock was, naturally, recorded in nickel (Fig. 3.31), where at $m = 20$ and $P_H = 24.7$ GPa, $\rho = 2.63$ g/cm^3. The shock compression energy in this case $E_H = 23.15$ kJ/g. Equal energy is achieved on the $m = 1$ Hugoniot at $P_H = 850$ GPa and $\rho_H = 16.9$ g/cm^3. Now let us examine the role of thermal components in these states. At $P_H = 24.7$ GPa the elastic ('cold') energy $E_c = 3.56$ kJ/g and $E_T =$

Shock compression of metals

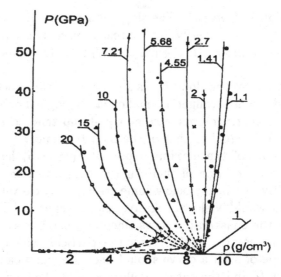

Fig. 3.31. $P-\rho$ curves for nickel, (laboratory measurements). Numbers by the curves are respective porosities.

$E_H - E_c = 19.59$ kJ/g, i.e., most of the shock compression energy is thermal. As concerns pressure, we have $P_c = -6.85$ GPa and $P_T = P_H - P_c = 31.6$ GPa. Compare these figures to respective parameters on the solid nickel Hugoniot. At $P_H = 850$ GPa we

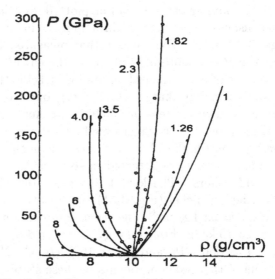

Fig. 3.32.   Same as in Fig. 3.31 for molybdenum.

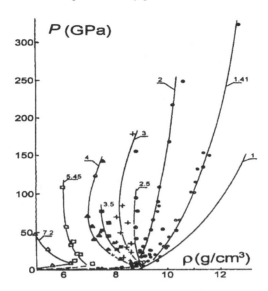

Fig. 3.33. Same as in Fig. 3.31 for copper.

have $P_c = 615$ GPa and $P_T = 235$ GPa, $E_c = 9.85$ kJ/g, and $E_T = 13.30$ kJ/g, i.e., the contribution of the elastic interaction between atoms is quite sizable. Thus in states of equal energies on Hugoniots of solid and porous metals, the energy is divided between the thermal and elastic components very differently.

Consider similar parameters on the $m = 7.2$ Hugoniot of porous copper (Fig. 3.33). At $P_H = 47$ GPa we have $\rho = 4.95$ g/cm^3 and $E_H = 14.0$ kJ/g. Equal energy on the solid copper Hugoniot is achieved at $P_H = 560$ GPa, when $\rho_{Cu} = 1.8\rho_{0,cr} = 16.1$ g/cm^3. The thermal energy on the porous Hugoniot at $m = 7.2$ at $P_H = 47$ GPa, $E_T = 14 - 1.5 = 12.5$ kJ/g ($E_c = 1.5$ kJ/g), i.e., most of its total energy is thermal. The thermal pressure in this state is $P_T = P_H - P_c = 47 - (-19) = 66$ GPa. On the solid copper Hugoniot at $P_H = 560$ GPa we have $P_c = 360$ GPa and $P_T = 200$ GPa, $E_T = 9.0$ kJ/g ($E_c = 5.0$ kJ/g and $E_H = 14$ kJ/g), i.e., in solid copper, as in nickel, the contribution of elastic ('cold') energy is also fairly large.

In $P - \rho$ coordinates (Figs. 3.31–3.33) Hugoniots of porous metals consist of two rectilinear sections, the low-pressure section is flat, and the high-pressure section is steep. The slope of the second section is a function of m. At $m < 2$ the derivative $(\mathrm{d}P/\mathrm{d}\rho)_H > 0$, at $2 < m < 3$ $(\mathrm{d}P/\mathrm{d}\rho)_H = \infty$ (vertical Hugoniots), at $m > 5$ the sign of the derivative is alternating, and at the highest m (Ni, Cu, Mo) $(\mathrm{d}P/\mathrm{d}\rho)_{\mathrm{H}} < 0$. Linear extrapolation of the second portions of porous Hugoniots across the cusp, characterized by the 'packing pressure', P_{p}, yields the density of a solid metal. The packing pressure P_{p} is a function of porosity m. The set of cusp points defines a certain enveloping line, which looks like the boundary of a two-phase region. But its shape cannot be accurately reconstructed because the measurement uncertainty is too large, especially on the first Hugoniot portion and at large m. Besides, an experimental error of 1% is difficult to achieve in this region because the operation of transducers is rather unstable under low pressure.

Now let us discuss the density uncertainty:

$$\Delta\rho = \rho \cdot \rho_0^{-1}(m\rho - \rho_0)\left(\left|\frac{\Delta U}{U}\right| + \left|\frac{\Delta D}{D}\right|\right).$$

If the uncertainty of D and U is standard (1–1.5%), the error in density, $\Delta\rho$, is controlled by those of m and ρ, and at a large m it may be quite considerable, although measurement errors compensate for each other at $\rho < \rho_0$. Nonetheless, given that numerous measurements obtained at various m in a wide range of P and ρ are self-consistent, we conclude that the shapes of high-pressure portions of Hugoniots are fairly accurate.

Let us consider investigations of metal compression, or, to be exact, test measurements, in the region of the $P - \rho - E$ diagram where neither theoretical nor experimental data are available. This region (Fig. 3.34) lies between the region of gaseous states (experiments with gases were reported in Refs. [57, 58], calculations are based on the Sakha equation) and the range of dynamic and static experiments with condensed matter, where calculations use the TFCK model. Needless to say how important is experiment in this region (it is shown by the hatched rectangle in Fig. 3.34) for verification of wide-range equations of state, which are designed to describe all states, including gases and highly compressed condensed states. The region defined by the hatched rectangle can be investigated using pressurized vapors of materials, in particular, metals. Unfortunately, such experiments are not feasible at present, primarily because of difficulties in fabricating the required gas samples. Other suggestions, such as the 'isochoric explosion' [59] could not be realized in due time because of difficulties in accurate measurements of neutron flux in underground nuclear tests. Another suggestion [56] was to investigate the states

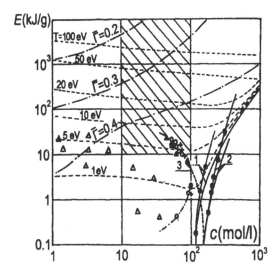

Fig. 3.34. Phase diagram: (1) Hugoniot and isothermal curve ($T = 0^0$ C) for gold; (2) similar curve for nickel; (3) Hugoniots of porous nickel: pluses $m = 10$; circles $m = 15$; solid triangles $m = 20$. Open triangles present Xe data from Refs. [57, 58].

within the 'white square' (Fig. 3.34) by measuring the compressibility of superporous samples with a porosity of more than ten and a starting density $\rho_{00} = 0.1 - 0.8$ g/cm^3. The first experiments were performed with nickel. We used very fine powder with particle dimensions of less than 200 Å manufactured for this experiment. X-ray diffraction measurements demonstrated that the powder was almost pure nonoxidized metal. The fabricated samples had densities $\rho_{00} = 0.44$, 0.59, and 0.89 g/cm^3 ($m = 20$, 15, and 10, respectively).

The Hugoniots of porous nickel are shown in Fig. 3.34. Beyond $P_H > 10$ GPa, they pass across the region of the $\rho - E - T$ diagram which had not been investigated previously, although only in its lower part. It is possible, however, to penetrate into the region of higher energy using facilities generating maximum pressure [30, 31]. Thus with nickel samples of very high porosity, one can investigate properties of metals in the region of parameters which was not accessible in previous experiments.

To conclude this section about the compression of porous metals, let us quote estimates of the Grüneisen coefficient Γ derived from Hugoniots of porous metals at maximum pressures realized in full-scale experiments. The Hugoniots of Cu, Fe, and W are approximately vertical lines close to the line described by the equation $\rho = \rho_{0,cr}$. As was noted above, in this case $\Gamma \approx \Gamma_{el}$ and approximately equals 0.5 for W and 0.7–0.9 for Cu and Fe. It seems that Γ is an individual parameter of a metal and is not determined by the number of the group to which the metal belongs.

3.6 Equations of state for metals

This section is a concise discussion of equations of state (EOS) based on various models, starting with the simplest ones (like Mie–Grüneisen) and ending with fairly complicated equations which describe all states of aggregation, including solid and gaseous, and have been proposed quite recently. Some of them have not been used in calculations for a long time and are discussed here only to present a comprehensive description of all approaches to the problem. All the discussed equations are presented in an analytical form and contain empirical parameters.

It is possible to calculate thermodynamic parameters of materials fairly accurately using the Monte-Carlo technique. This calculation demands only the interaction potential between atoms. But it is virtually impossible to accurately determine this potential, and Monte-Carlo calculations are rather complicated and cumbersome.

For this reason, semi-empirical models are often used in practice, i.e., models describing thermodynamic properties of materials and using various equations of state based on simple concepts of the material structure and behavior under external fields. These equations, which relate thermodynamic parameters, such as pressure, density, and energy (temperature), are brought to coincidence with experimental data through varying empirical (fitting) parameters derived from experimental data.

These empirical data are usually thermodynamic parameters of materials under normal conditions ($P = 0$, $T = 20^0$ C), and static and dynamic measurements of compression, adiabatic expansion, etc. Measurements of dynamic compression are especially important because they are more accurate and performed over wider ranges of thermodynamic parameters of materials in various phase states than measurements of other types.

Depending on the problem to be solved, the employed equation of state may be more or less complicated and contains more or fewer empirical parameters derived from experimental data.

The stricter the requirements the equation of state, the more fitting parameters are, usually, introduced into the equation. Most of the equations which are called wide-range (i.e., they are designed to describe all states of a material and yield correct asymptotics of thermodynamic parameters in limiting cases) contain two or three dozen, or even more empirical constants used as fitting parameters. Such equations are, naturally, rather cumbersome, and a calculation of thermodynamic functions using such equations is not an easy job.

Therefore, it is natural that an equation with fewer fitting parameters, which makes the calculation of thermodynamic parameters easier, is preferable, provided that it adequately describes experimental data. A fairly simple equation may be good if the researcher has 'guessed' underlying physical processes in the material under various conditions when selecting the form of the equation.

In this section we shall briefly discuss some equations of state, including both the simplest forms and more complicated wide-range equations. Most of them have been used at different times to solve problems of continuous medium motion, including calculations of design parameters of structures in the gas-dynamic stage of their modelling. Some of these equations are still important for understanding physical processes in materials.

In the simplest form, states of materials are described by the Mie–Grüneisen equations:

$$P = P_c + P_T = P_c + \Gamma c_v T / v, \qquad E = E_c + E_T = E_c + c_v T. \quad (3.1)$$

In the scheme which we use most often, the total pressure and energy are presented as sums of two components, namely the potential ('cold') part due to the elastic interaction between atoms at zero temperature ($T = 0$ K) and the thermal part taking into account the kinetic energy due to thermal motion.

By excluding the temperature from Eq. (3.1), we obtain the caloric equation

$$P - P_c = \Gamma(E - E_c)/v. \quad (3.2)$$

Here v is the specific volume, P is pressure, E is energy, and Γ is the Grüneisen parameter.

Taking $P = P_H$ and $E = E_H$ (the index H denotes a Hugoniot adiabat) and using the relation $P_c = -dE_c/dv$, we relate the parameters P_H and v_H of the Hugoniot to P_c and E_c (these relations are called Hugoniot equations of state):

$$E_c = \frac{1}{(h-1)v^{\frac{2}{h-1}}} \int_{v_0}^{v} P_H(v) \left(\frac{v_0}{v} - h \right) v^{\frac{2}{h-1}} dv$$

$$P_c = \frac{2}{(h-1)^2 v^{\frac{h+1}{h-1}}} \int_{v_0}^{v} P_H(v) \left(\frac{v_0}{v} - h \right) v^{\frac{2}{h-1}} dv -$$

$$\frac{1}{h-1} P_H(v) \left(\frac{v_0}{v} - h \right). \tag{3.3}$$

Here $h = 1 + 2/\Gamma$ is the shock compressibility limit.

Equation (3.3), which is determined by the selected parameters Γ (or h) and $P_H(v)$, is valid in the area on the $v - P$ diagram defined by the isothermal curve $P_c(v)$ below it and the Hugoniot $P_c(v)$ above.

In one version of Eq. (3.1), the function $P_c(v)$ is defined in the most simple presentation in the form of a power function $P_c = s_0^2(v_0/v - 1)/v_0 n$, where s_0 and n are empirical parameters and v_0 is the initial volume, $E_c = -\int_{v_0}^{v} P_c(v)dv$, and the Hugoniot equation is presented in the form

$$P_H = \frac{s_0^2}{v_0 n} \frac{[h - (n+1)/(n-1)](v_0/v)^n + 2nv_0/(n-1)v - (h+1)}{h - v_0/v}.$$

These equations cannot describe experimental data with sufficient accuracy, especially at high dynamic parameters. Moreover, there is a fundamental limitation on this model since it can be used only in the range of parameters close to the Hugoniot adiabat and characterized by relatively small temperatures (2–3 eV), at which the effect of thermally excited electrons is small.

At higher temperatures their effect should necessarily be taken into account, which was done by Al'tshuler et al. [60], who proposed the equations of state including, unlike the Mie–Grüneisen equations, the higher terms due to electron thermal components in pressure and energy (P_{Te} and E_{Te}):

$$P = P_c + P_{Tl} + P_{Te} = P_c + \Gamma_l(v)c_{vl}(T - \hat{T})/v + 0.5\Gamma_e\beta T^2/v$$

$$E = E_c + E_{Tl} + E_{Te} = E_c + c_{vl}(T - \hat{T}) + 0.5\beta T^2, \tag{3.4}$$

where P_{Tl} is the pressure due to lattice, E_{Tl} is the lattice energy, $\Gamma_l(v)$ is the Grüneisen coefficient, $\hat{T} = T_0 - E_0/c_{vl}$, E_0 is the initial energy at normal conditions, c_{vl} is the lattice specific heat, $P_{\text{Te}} = \Gamma_e E_{\text{Te}}/v = \Gamma_e \beta T^2/2v$, Γ_e is the electron analogue of the Grüneisen coefficient, β is the factor of the electron specific heat, $\beta = \beta_0 \exp(\int_{v_0}^v \Gamma_e dv/v)$. At $\Gamma_e = 0.5$ we obtain $\beta = \beta_0(v/v_0)^{1/2}$. The latter statement is valid only at relatively low temperatures, when the electron gas is approximately degenerate.

Note that techniques for constructing EOS are usually based on some general assumptions:

– any phase and electronic transitions in the solid phase, and chemical reactions due to shock are ignored;

– effects of changes in the material strength are also ignored, and dynamic compression is assumed to take place under conditions of thermodynamic equilibrium.

In Eq. (3.4) we have postulated that the lattice specific heat is constant with temperature and density, hence the Grüneisen parameter, Γ_l, is also independent of the temperature. Having adopted this assumption, we ignore the anharmonic components of the interatomic potential of the lattice, therefore our equation is inaccurate under an intense heating due to a shock.

The next step in improving the accuracy of equations of state was made by Al'tshuler et al. [61] and Kormer et al. [53]. In the first of these publications, the authors took into account the temperature dependence of c_v and Γ_l using the theory of 'free volume'. The temperature dependence was introduced through the formulas $E_{\text{Tl}} = \overline{c_v}(T) \cdot T$ and $P_{\text{Tl}} = \overline{\Gamma_l}(v,T)\overline{c_v}(T)/V$, where $\overline{c_v}(T)$ is the lattice thermal capacity averaged over the temperature range between 0 and T, and $\overline{\Gamma_l}(v,T) = \Gamma_l(v,0) - \Delta\Gamma(v,T)$. The anharmonicity of thermal oscillations of atoms was introduced according to the 'free volume' theory [62].

More information was derived using an interpolation equation with variable thermal capacity and Grüneisen parameter, which was proposed by Kormer et al. [53] and later used by many researchers.

The transition of the lattice thermal capacity, c_v, from $3R$ for a 'cold' solid at a temperature of several Debye temperatures Θ to $3R/2$ at $T \gg \Theta$, i.e., in the state of very hot liquid or dense gas, and the transformation of the analogue of the Grüneisen parameter, λ_l, from Γ_0 under normal conditions to $2/3$ at $T \to \infty$ are described by introducing the parameter $Z = lRT/s_c^2$, where l is the empirical constant, R is the gas constant, and s_c is the speed of sound on the isothermal curve at $T = 0$ K:

$$c_v = \frac{3}{2}R\left[1 + (1 + Z)^{-2}\right] \qquad \lambda_l = 2(3\Gamma_l + Z)/3(2 + Z)$$

$$P_{\text{Tl}} = (3\Gamma_l + Z)RT/(1 + Z)v \quad E_{\text{Tl}} = \frac{3}{2}(2 + Z)RT/(1 + Z). \quad (3.5)$$

Here the Grüneisen parameter, Γ_l, is related to s_c through the equation

$$\Gamma_l = \frac{1}{3} + \frac{d\ln s_c}{d\ln\rho}, \qquad (\rho = 1/v).$$

It follows from Eq. (3.5) that in the limiting cases P_{Tl} and E_{Tl} are expressed by the Mie–Grüneisen equation (solid state, $Z = 0$) with $c_v = 3R$ and $\lambda_l = \Gamma_l$ and by the gas equation of state, $Pv = RT$, $c_v = 3R/2$, and $\lambda_l = 2/3$ ($Z \to \infty$). At intermediate Z the effect of temperature and density is taken into account through the parameter Z.

The elastic interaction of atoms in Ref. [53] is described by the equation

$$P_c = \sum_{i=1}^{7} a_i(v_k/v)^{1+i/3}, \qquad (3.6)$$

where v_k is the volume at $T = 0$ K and $P = 0$. The first three coefficients in this series are derived [53] from literature data about ρ_0, Γ_0, and compressibility factor, two coefficients are derived from P_{H} determined by Hugoniots of bulk materials, and two coefficients are derived from calculations of P_c and its derivative P_c' by the TFCK model [34].

The electronic component in the equations was described by a function approximating the data reported by Latter [63]:

$$E_{\text{Te}} = \frac{b^2}{\beta}\ln\cosh\frac{\beta T}{b}, \qquad (3.7)$$

where b is the fitting parameter. The pressure is related to the energy through the equation $P_{\text{Te}} = \Gamma_e E_{\text{Te}}/v$.

The equation of state is derived from Eqs. (3.5–3.7). In this approach the transition from the solid equation of state to the effective liquid equation is described through the variation of the parameter Z. The algorithm for calculating the melting curve was given by Urlin [64]. He described the liquid state by adding to the equation for the solid phase an empirical function of temperature and density. The constants of the additional summand were derived by fitting the calculation to the experimentally determined melting curve, and the theoretical melting curve was derived by minimizing the Gibbs potential.

The discussed equations from Ref. [53] contain sixteen constants. Some of them have clear physical sense, such as the initial density, speed of sound, Grüneisen parameter, gas constant, and coefficient of the electronic thermal capacity, the rest are fitting parameters. Fits of calculated curves to experimental data are shown in Figs. 3.35, 3.36 demonstrating measurements of copper and nickel. You can see that most Hugoniots are satisfactorily described using the proposed equation of state.

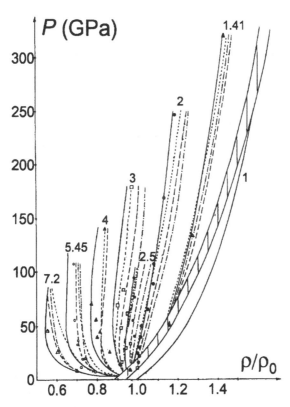

Fig. 3.35. Comparison between calculations and experimental data for Cu. Calculations were done using equations from Ref. [75] (solid line), Ref. [70] (dashed line), Ref. [65] (dash-dotted line), Ref. [53] (dotted line). Numbers by the curves are respective porosities. The melting region is hatched.

Shock compression of metals

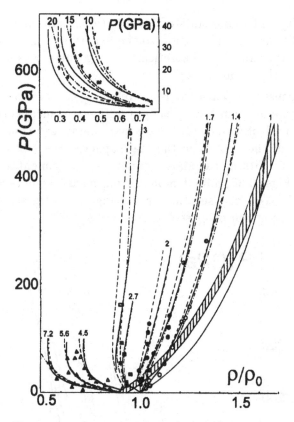

Fig. 3.36. Comparison between calculations and experimental data for Ni. Calculations were done using equations from Ref. [75] (solid line), Ref. [67] (dashed line), Ref. [53] (dash-dotted line). Numbers at the curves are respective porosities. The melting region is hatched.

Sapozhnikov and Pershina [65] and Al'tshuler et al. [59] used different interpolations. In the first of these two publications, the equation of state was modified to describe expansion of initially compressed materials to low-density states ($\rho < \rho_0$), and a more simple and convenient form of the function E_{Te} was selected.

The cold components were expressed, as in Ref. [61], in terms of the Born–Meyer potential, which included an additional function to join smoothly the cold isotherm to the calculation by the Thomas–Fermi model [34]. Besides, the constants in the potential (elastic) components [61] were modified so that the calculated binding energy at $v_0/v < 1$ should conform to experimental data.

The thermal components were obtained by generalizing the

equations of Ref. [53] to describe vapor states with $\Gamma_l(\rho)$ expressed as a simple power function in the range $\rho < \rho_0$. As in Ref. [53], the equations of state are reduced to the Mie–Grüneisen equation and gas equation in the limiting cases.

The electronic component of thermal energy is expressed at low temperatures in the form

$$E_{Te} = \frac{\beta(\sigma)}{2} \cdot \frac{T_F T^2}{T_F + T(\sigma_* + \sigma)^{-\Gamma_e}},$$

where Γ_e, T_F, and σ_* are the empiric constants, $\beta(\sigma) = \beta_0(\sigma_* + \sigma)^{-\Gamma_e}$, to adequately describe experimental data at low temperatures in the form $E_{Te} = \beta T^2/2$, and at high temperatures ($T \gg T_F$, where T_F is an equivalent of the Fermi temperature) $E_{Te} = \beta T_F T/2$. The full equation contains about twenty fitting parameters, most of which are varied to fit calculations to experimental data.

Figures 3.35, 3.36 show as an example curves calculated by this equation of state from Ref. [65] fitted to measurements of copper and nickel. As in the previous case, the agreement between calculations and measurements is satisfactory.

The version of the interpolation equation proposed by Al'tshuler et al. [59] is based on the concepts of Refs. [53, 65]. The cold and nuclear thermal components are the same as in Ref. [53], namely, the power series for cold components and the interpolating equation for the nuclear thermal components with the parameter $Z = lRT/s_c^2$. The electronic component at relatively low temperatures is described by a quadratic function of temperature, and at high temperatures ($T > 5$ eV) by the function proposed in Ref. [65]. The choice of the interpolating curve for P_c is, in a sense, inconsistent because parameters of simple metals (except aluminum) are calculated by the Thomas–Fermi model [63] and those of transition metals (Mo and Fe) and Al by the TFCK model [34].

This equation of state also contains a lot of empirical constants (thirty-one), twenty-five of which are fitting parameters determined by fitting TF and TFCK calculations to experimental data.

Glushak et al. [66] calculated $P_c(v)$ using experimental data, including those from shock compression measurements, in the form of a table, then they approximated the curve by a two-component power function.

The Grüneisen parameter, Γ_l, as a function of density was also derived from the best fit of calculations to experimental data. Note, however, that in all cases this form of the function was based on some theoretical model. The analogue of the Grüneisen parameter for electrons in Ref. [66] is a function of density. The equation contains twenty-six constants of which twenty-three are fitting parameters.

Al'tshuler et al. [67] proposed an equation containing cold, thermal nuclear, and electronic summands in the traditional form. The cold pressure at $\sigma_k = v_k/v$, where v_k is the specific volume at $P = 0$, $T = 0$ K, is defined, as in Ref. [68], by a sum similar to that in Eq. (3.6):

$$P_c(\sigma_k) = \sigma_k^{4/3} \sum_{i=1}^{6} a_i(\sigma_k^{i/3} - 1),$$

and at $\sigma_k < 1$ by the polynomial

$$P_c(\sigma_k) = A(\sigma_k^m - 1) + B(\sigma_k^n - 1).$$

The two branches of the curve are smoothly joined at the crossing point with the abscissa ($P = 0$), and the asymptotics in the regions of high compression and expansion are correct.

The lattice component of the potential is divided into two parts: the first one characterizes the solid state, and the second the liquid state. The free energy of the first part is given by

$$F(\sigma, T) = 3RT \ln [\theta(\sigma)/T], \qquad (3.8)$$

where $\theta(\sigma)$ is a certain 'effective' frequency determined at a fixed $P_c(\sigma)$ via the Grüneisen parameter: $\Gamma(\sigma) = \mathrm{d}\ln\theta/\mathrm{d}\ln\sigma$. The latter is expressed by an interpolating formula which yields the asymptotic $\Gamma = 2/3$ at $\sigma = 0$ and $\sigma \to \infty$. The frequency θ is derived by integrating the function Γ, whereby the lattice component of the solid phase thermal potential is determined using Eq. (3.8).

The liquid phase potential also contains two summands:

$$F_L = F_{L_1} + F_{L_2}.$$

The first summand, $F_{L_1}(v, T)$, is due to the vibrational spectrum and thermal capacity of high-temperature states ($T \gg T_{\text{melt}}$), the second summand describes effects of melting. The first component, F_{L_1}, is expressed by a function similar to that in Eq. (3.8), but it should provide a transition of the specific heat from $3R$ to $3/2R$ for the ideal gas (in this transition Γ changes to $2/3$):

$$F_{L_1} = 3RT\varphi(T,\sigma)\ln\frac{\theta_{L_1}(\sigma,T)}{T},$$

where $\varphi(T,\sigma)$ is the interpolation function introduced to describe the transition to the gas specific heat. $\theta_{L_1}(\sigma,T)$ is described by an appropriate function, in which the frequency depending only on σ is included in a separate summand. It is determined by integrating $\Gamma(\sigma)$ in the liquid phase. The expression for F_{L_2} is constructed so as to correctly describe the experimental curve with the density jump and temperature increase at the melting point.

The electronic component of the potential $F_e(\sigma,T)$ is expressed by a function which includes generalized analogues of the electronic thermal capacity factor, Grüneisen parameter $\Gamma_e(\sigma,T)$, and electronic specific heat $c_e(\sigma,T)$.

In the condensed phase, at relatively low temperatures, $F_e(\sigma,T) \sim T^2$, at high temperatures it is an almost linear function of temperature.

The parameter Γ_e is defined by a function which provides a smooth transition from $\Gamma_e = \Gamma_0$ ($T = 0$ K) to $\Gamma_e = 2/3$ at $T \to \infty$ ($\sigma = 0$ and $\sigma = \infty$).

The empirical function for the coefficient of the electronic specific heat $\beta(T)$ is designed to have limiting values β_0 at low temperatures and β_i at high temperatures. The function $c_e(\sigma,T)$ satisfies similar conditions: it should be equal to the parameter of an ideal electronic gas at $T \to \infty$ and describe properties of ionized matter in the plasma state.

The combination of all these expressions for potential and thermal components yields the wide-range equation of state for metals, which is designed to correctly describe thermodynamic states of metals throughout their phase diagram. Unfortunately, EOS is very cumbersome. Its thermal summands contain complex logarithmic, exponential, and trigonometric functions. The equation contains forty-three empirical constants, twenty-seven of which are fitting parameters. They are derived, as usual, from both static and dynamic measurements, and also from theoretical calculations of states under extreme conditions. Later this equation underwent minor modifications, new versions of it were published, for example, by Bushman and Fortov [69].

Some progress in the development of semi-empirical equations of state with variable specific heats of nuclei and electrons resulted from the work by Glushak et al. [70]. The resulting equation of state ROSA was a generalization of equations described in Refs. [53, 59, 65] and had two specific features:

1. The function $\Gamma_l(\sigma)$ in the studied range of compression (including $\sigma < 1$) was derived from experimental data without using specific models of matter, unlike Refs. [53, 65].

2. The electronic Grüneisen parameter Γ_e was a function of density, $\Gamma_e = \Gamma_e(\sigma)$.

The equation was presented in the traditional form with summands corresponding to cold, thermal lattice, and thermal electronic components. The parameter P_c was described by a power function:

$$P_c = a_i \sigma^{n_i} + b_i \sigma^{m_i}, \tag{3.9}$$

where i is the number of the interval in which the function is approximated, $\sigma = \rho/\rho_0$, $E_c = \int P_c(\rho)/\rho^2 d\rho$.

The lattice thermal functions are

$$E_{\text{Tl}} = \int c_v(\sigma, T) d\tau; \qquad P_{\text{Tl}} = \Gamma(\sigma, T) \rho E_{\text{Tl}}. \tag{3.10}$$

In these expressions

$$c_v(\sigma, T) = \frac{c_{v0}}{2}\left[1 + \frac{\psi^2(\sigma)}{(\psi(\sigma) + T)^2}\right]; \quad \Gamma_l(\sigma, T) = \frac{2}{3}\frac{3\psi(\sigma)\Gamma_l(\sigma) + T}{2\psi(\sigma) + T}.$$

Here $\psi(\sigma)$ is a function of relative compression. One can see that in limiting cases correct asymptotics can be derived from these equations, namely $c_{v0} \to 3R/A$ (A is the atomic number) and $\Gamma_l(\sigma, T) \to \Gamma_l(\sigma)$ at small temperatures for the solid phase, and $c_v(\sigma, T) \to 3R/2A$ and $\Gamma_l \to 2/3$ for ideal gas.

At intermediate temperatures $c_v(\sigma, T)$ and $\Gamma_l(\sigma, T)$ are determined by the function $\psi(\sigma)$, which is, in its turn, derived from the function $\Gamma_l(\sigma)$ expressed in the form of a power function.

The thermal electronic functions P_{Te} and E_{Te} were expressed in a form similar to that adopted in Ref. [65], the only difference was that Γ_e was a function of density. As in Ref. [65], the electronic thermal capacity had correct asymptotics:

$$c_{ve} = \beta T \text{ at } T \ll T_{\text{F}}; \qquad c_{ve} = \frac{1}{2}\beta_0 T_{\text{F}} \text{ at } T \gg T_{\text{F}}.$$

The equation is fully determined, given a set of physical constants (ρ_0, c_v, β_0, Γ_0, etc.) and also twenty-three fitting parameters whose values are derived from experimental data. Although the model used to derive the ROSA equation of state was designed carefully, the melting of the material was ignored, which significantly degrades the scientific value of this equation. The ROSA equation of state [70] was the last in the series of EOS composed by separating the total free energy into three components, the first of which determines the elastic energy of atoms in the lattice at $T = 0$ K, the second one the contribution from thermal oscillations of atoms, and the third one thermal motion of excited electrons.

Although different components in the equations of state considered above were obtained using different approaches, note that the resulting isothermal functions for P_c were quite close. Isothermal curves for copper obtained from different equations of state are compared in Fig. 3.37. In the pressure range of up to 800 GPa they are very close to each other: the distance between the aver-

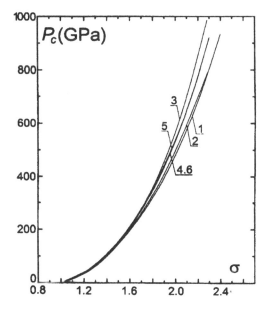

Fig. 3.37. Comparison between isothermal curves of cold compression derived from different equations of state: (1) Ref. [67]; (2) Ref. [75]; (3) Ref. [70]; (4) Ref. [65]; (5) Ref. [59]; (6) Ref. [53].

age curve and the extreme curves along the σ axis is within 0.07. The deviation of curves calculated for other metals from different EOS are similarly close. But this cannot be said for thermal components: calculations by different EOS yield very different results.

A different approach to the equation of state was used by Zubarev et al. [71, 72]. It was expressed in the following form:

$$E(P, v) - E_1(P) = \frac{P}{\eta(P)}[v - v_1(P)] \tag{3.11}$$

with the proportionality factor $\eta(P) = P\left(\frac{\partial v}{\partial E}\right)_P$. This form was chosen for the following reasons. The Mie–Grüneisen equation contains the proportionality factor Γ. Its average numerical value,

$$\overline{\Gamma} = \frac{\Delta P_T}{\rho \Delta E_T},$$

is usually determined by comparing porous Hugoniots plotted in the $P - v$ plane. In this comparison the cold energy and pressure (E_c and P_c) at a given v are taken equal for all Hugoniots (porous or bulk with $\rho_{0,\mathrm{cr}}$), and the pressure difference $P_{H_1} - P_{H_2} = \Delta P_T$ is determined by the difference between thermal energies $\Delta E_T = E_{H_1} - E_{H_2}$. But in most cases porous Hugoniots are located in the $P - v$ plane so that it is difficult to determine P_H at crossing points of two Hugoniots at a given v with sufficient accuracy (e.g., on steep sections of Hugoniots). This difficulty can be overcome by using Eq. (3.11) with the parameter $\eta(P)$. Its numerical values derived from Hugoniots of porous materials yield η as a function of other parameters, except pressure. In Eq. (3.11) $E_1(P)$ and $v_1(P)$ are the parameters of a reference curve, which is usually either a solid Hugoniot or an isentrope originating at $P = 0$, $T = 300$ K. In the first case the Hugoniot is expressed in the parametric form as $v_1 = v_0[1 - U/(s_0 + \lambda U)]$, $P = U(s_0 + \lambda U)/v_0$) (here we assume that the Hugoniot in $D - U$ coordinates is rectilinear, $D = s_0 + \lambda U$).

In combination with the energy conservation equation, $E_1(P) - E_0 = P[v_0 - v_1(P)]/2$, these expressions determine $v_1(P)$ and $E_1(P)$ in Eq. (3.11). Thermodynamic parameters which determine positions of double-compression Hugoniots and porous Hugoniots are derived from Eq. (3.11). The isentropic condition $dE = -Pdv$ in combination with Eq. (3.11) determines the differential equation for isentropes:

$$\frac{dH}{dP} - \frac{\eta H}{P(1 + \eta)} = \frac{2 + \eta}{2(1 + \eta)}v_1(P) - \frac{\eta}{1 + \eta}\left(\frac{E_0}{P} + \frac{v_0}{2}\right),$$

where $H = E + Pv$ is the enthalpy, E_0 and v_0 are the energy and volume in the initial state, $v_1(P)$ is the specific volume on the reference Hugoniot of water, and the temperature is determined by the equation

$$T = T_0 \exp \left[\int_{P_0}^{P} \frac{\eta(P)}{1 + \eta(P)} \frac{dP}{P} \right].$$

The function $\eta(P)$ is expressed in the analytical form with a few fitting parameters (usually no more than four). The disadvantage of the equation is its limited range of application, which is largely determined by the shock compression available in experimental facilities. The equation, however, can be modified by introducing other parameters so that it could describe states under extreme conditions.

The equation of state proposed by Medvedev [73, 75] and Kopyshev [74] is a very special case since it describes the processes taking place in compression (expansion) of materials in the solid, liquid (melted), and gaseous states differently from equations of other types. It uses the simplest Mie–Grüneisen model (Eq. (3.1)) to describe the solid state, and it gradually transforms to the modified Van-der-Waals model of liquids. Let us recall that at moderate temperatures the Van-der-Waals equation for liquids in its simplest form is presented as

$$P(v, T) = \frac{RT}{v - v_c'} - \frac{A}{v^2} \qquad (3.12)$$

with two empirical constants: A is the constant determined by attracting forces, and v_c' is the co-volume, i.e., the smallest volume occupied by densely packed liquid molecules, which are considered as incompressible balls.

The first summand on the right of Eq. (3.12) is the pressure due to repulsion, the second summand is due to attraction. The major difficulty in handling this equation is due to the introduction of co-volume: since molecules are assumed to be incompressible, the system volume has a lower limit: $v > v_c'$. Medvedev [75] assumed that v_c' may be a function of pressure, but independent of temperature. Thus $v_c' = v_c'(P_{\text{rep}})$, and this function is varied to fit calculations to experimental data. Thus the main flaw of the 'co-volume' Van-der-Waals equation was eliminated. Since the temperature is not specified, the equation of state is determined by the function $v_c'(P_{\text{rep}})$ at all temperatures.

Then the function $v_c'(P_{\text{rep}})$ is replaced with the inverse function $P_{\text{rep}}(v_c')$ [75], and the system is described by the following

Shock compression of metals

equations in the variables T and v_c':

$$v = v_c' + \frac{RT}{P_{rep}(v_c')} \tag{3.13}$$

$$P = P_{rep}(v_c') + P_{attr}(v). \tag{3.14}$$

The energy in this model is expressed as

$$E = E_0 + E_{rep}(v_c') + E_{attr}(v) + \frac{3}{2}RT, \tag{3.15}$$

where

$$E_{rep}(v_c') = -\int_{-\infty}^{v_c'} P_{rep}(v_c')dv_c'$$

$$E_{attr}(v) = -\int_{-\infty}^{v} P_{attr}(v)dv.$$

Hence the free energy and entropy are derived using the relations $F = E - TS$ and $S = -\left(\frac{\partial F}{\partial T}\right)_v$. Equations (3.14–3.15) define the equation of state through the parameter v_c'. This equation accounts for the evaporation of the liquid, its compression at low temperatures, and reduces to the ideal gas equation at high temperatures. The empirical functions $P_{rep}(v_c')$ and $P_{attr}(v)$ are constructed using simple relations with few constants derived from experimental data. In Ref. [73] the co-volume was defined as $v_c = v_c'$, which was derived from the function $P_c(v)$. Thus the curve of cold compression determined the co-volume. In Ref. [75] $P_{attr}(v) = -A\,(v/v_k)^{-n}$, $P_{rep}(v_c') = A\exp[const\,(1 - v_c'/v_k)]$, where A, n, and v_k are constants. The curve of the solid state is smoothly joined to that of the liquid using the criterion similar to that proposed by Lindeman.

In order to describe high-temperature states, the model is generalized to include thermal ionization [76]. The concentrations of ions, electrons, and atoms are calculated by solving equations of ionization equilibrium similar to the Sakha equation and yielding the Sakha solution in the high-temperature limit.

Parameters of the model were derived from thermodynamic characteristics under normal conditions, literature data about ions, parameters of melting at atmospheric pressure, and parameters of the Hugoniot originating from the point with the crystalline initial density. In addition to its clear physical meaning, the advantage of the model is that it contains only seven empirical constants, of which only four are fitting parameters.

Curves calculated by the latter model and fitted to experimental data are shown in Figs. 3.35, 3.36. You can see that at a porosity of up to ten, the agreement between calculations and measurements is satisfactory. At a higher porosity the deviation of calculations from experimental data is considerably larger. The reason for this is still unclear, probably, a compressed metal with an initial porosity $m > 15$ (the final temperature more than $5 \cdot 10^4$ K) is transferred to a state with a gap in its spectrum that results in a loss of its conductivity. None of the equations discussed in this section and other empirical equations take account of this effect.

3.7 Compressibility of initially melted and supercooled metals

So far we have discussed metals at normal temperatures in their initial states. This section presents investigations of metals heated to several hundred degrees, including low-melt metals, which are liquid at such temperatures. The interest in these experiments is driven by two incentives. Firstly, one can compare Hugoniots of low-melt metals with different initial states, namely, solid at normal temperature and liquid at $400 - 450^0$ C. This comparison was expected to yield additional information about melting processes on the shock front. Secondly, our intention was to develop a technique for detecting parameters of shock in heated metals with a view to tuning the position of a Hugoniot on a $P - T$ diagram and even to driving it across the line of phase transition. Experiments with bismuth samples cooled to a temperature $T = -185^0$ C were designed for the same purpose.

The experiments were performed using the traditional impedance-matching technique. Tested samples were placed downstream of a copper shield. They were heated by a nichrome wire heater or cooled by immersing the entire device into a vessel with liquid nitrogen. The temperature was measured by thermocouples. The electric sensors were metal rods insulated from melted metal. Although the shield was thermally insulated from a sample, its temperature also increased up to $T \simeq 400^0$ C when the sample was heated. For this reason, we previously recorded the Hugoniot of copper with an initial temperature $T_0 = 400^0$ C. We found that the Hugoniot of heated copper plotted in $D - U$ coordinates coincided with that recorded with 'normal' initial conditions within

Shock compression of metals

the 1% measurement uncertainty. But the Hugoniot plotted in $P - U$ coordinates was notably different because of the different initial density of heated copper ($\rho = 8.74\,\mathrm{g/cm^3}$), which was taken into account in processing experimental data.

We studied the shock compression of melted Bi, Cd, Pb, Sn, and Zn at an initial temperature of 400^0 C (the starting temperature of Zn was $+450^0$ C). Hugoniots of Bi with an initial temperature of -185^0 C were also recorded.

The experimental data are given in Fig. 3.38, where they are compared to 'normal' Hugoniots ($P_0 = 0$, $T_0 = 20^0$ C, $\rho = \rho_0$).

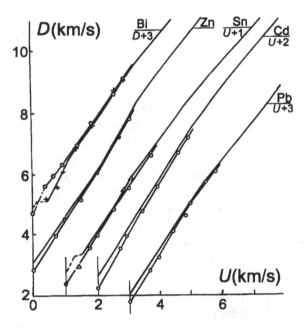

Fig. 3.38. Comparison of Hugoniots of metals with initial states at $T = 20^0$ C and $T = 400\text{--}450^0$ C ('normal' and 'hot' Hugoniots). Experiments with melted metals: circles, triangles; 'cold' bismuth: pluses.

By and large, the 'cold' and 'hot' Hugoniots are very similar and originate from the points corresponding to the speed of sound in the materials. The 'hot' Hugoniots of all metals, except Bi, lie below 'normal' Hugoniots and their slopes, D'_U, on low-velocity portions are larger than those of 'normal' $D - U$ Hugoniots. At velocities higher than some critical value, D_{cr}, the differences between the Hugoniots are either zero (Bi, Zn) or so small that they are within the measurement uncertainty. At the critical point the slope of the 'hot' $D - U$ Hugoniot also changes and becomes equal to that of the 'normal' Hugoniot.

This phenomenon can be interpreted in terms of metal melting on the 'normal' Hugoniot at D_{cr}. The critical shock velocities are: 3.5 km/s (Bi), 4.6 km/s (Zn), 4.6 km/s (Sn), and 4.2 km/s (Cd). The critical velocity in Pb was not measured $(D > 3.5$ km/s).

Dynamic measurements of various metals [28, 46, 47] were analyzed more thoroughly in terms of this concept of melting. Figure 3.39 shows shock measurements of bismuth and tin. The tin Hugoniot can be approximated in the 3.5 km/s $< D < 6$ km/s

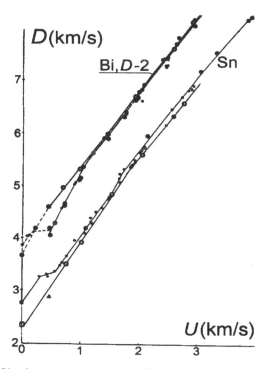

Fig. 3.39. Similar curves for bismuth and tin plotted through experimental data for 'normal' (dots, filled circles), 'hot' (open circles) and 'cold' (opluses for bismuth) Hugoniots.

interval by three rectilinear portions of different slopes. The critical velocity for tin also lies in this interval, $D_{cr} = 4.6$ km/s. It is highly plausible that the change in the Hugoniot slope is caused by tin melting on the 'normal' Hugoniot. The proximity of 'hot' and 'cold' Hugoniots at this point and equality of their slopes indicate that the phase states on both curves are identical. This is also

Shock compression of metals

illustrated by the comparison of Hugoniots of 'hot', 'normal', and 'cold' bismuth. At the point $D_{cr} = 3.5$ km/s, $U_{cr} = 1.2$ km/s all three Hugoniots have close positions and slopes, and, given that the 'hot' Hugoniot describes the liquid phase, this indicates that bismuth is liquid on all these curves beyond the critical point.

The 'hot' and 'normal' Hugoniots in $P - \rho$ coordinates are compared in Fig. 3.40. The thick lines mark the plausible pressure intervals in which the metals are melted. These intervals con-

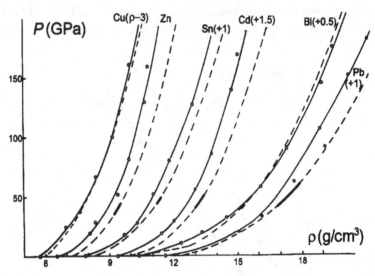

Fig. 3.40. Comparison of 'normal' (dashed line) and 'hot' (solid line) Hugoniots plotted on the $P - \rho$ plane. The highlighted section shows the region of melting on the 'normal' Hugoniot.

form to the estimates of melting conditions derived from modified equations of state.

To end this section, we must emphasize that the conclusions about melting of metals under shock compression from available experimental data are very approximate. Ultimate conclusions could be made only after high-precision measurements of shock kinematic parameters in a quasi-continuous mode.

4

Compression of
metal compounds

4.1 Shock compression of metal alloys

To this point, we have discussed the shock compression of almost pure elementary metals. But in most cases alloys rather than pure metals are used for technical needs. Since some units and devices of nuclear facilities operate under high shock loads, dynamic parameters of materials of these units are necessary to predict their behavior under emergency conditions. Sometimes shock parameters of specific materials and units are measured, especially those of critical components of nuclear reactors and other accident-prone facilities. But, given the abundance of alloys used in various technologies, it is impossible to study all of them, therefore techniques for calculating their parameters using measurements of a few materials are badly needed. Some techniques are based on the principle of additivity, i.e., specific volumes of alloy components are added at a given pressure. It is known that the additivity method is, strictly speaking, applicable to the processes of isothermal compression, when not only the pressures, but temperatures in different components are also equal. It turns out, however, that in most cases the straightforward addition of specific volumes of different Hugoniots also yields satisfactory results, which makes calculations notably easier. In this case the alloy specific volume, $V_{\mathrm{al}}(P)$, at a given pressure, P, is made equal to the sum of specific volumes of its components derived from their Hugoniots and multiplied by their contents:

$$V_{\mathrm{al}}(P) = \sum_i \alpha_i V_i(P),$$

95

where α_i is the weight fraction of the i-th component, $V_i(P)$ is its specific volume, and $\sum_i \alpha_i = 1$. In a two-component system $V_{al}(P) = \alpha_1 V_1(P) + (1 - \alpha_1)V_2(P)$.

This method was tested by analyzing measurements of about forty materials [77], the amount of experimental data being quite considerable. This set included alloys of all three main groups, namely, chemical compounds, mechanical mixtures, and solid solutions. Alloys with more than two components and alloys of iron and aluminum were treated as separate groups.

In the derivation of additive curves from Hugoniots of individual metals, we used appropriate analytical expressions [32]. The calculated curves of $P(\rho)$ are compared to experimental data in Figs. 4.1, 4.2. You can see that, by and large, the agree-

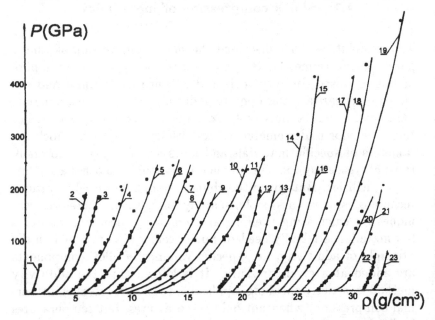

Fig. 4.1. Calculated 'additive' Hugoniots of metal alloys compared to experimental data: (1) Be 68%, Al 32%; (2) Fe 90%, Ni 10%; (3) Fe 74%, Ni 26%; (4) Cu 74.5%, W 25.5%; (5) Cu 45%, W 55%; (6) Cu 32%, W 68%; (7) Cu 24%, W 76%; (8) Au 90.7%, Ge 9.3%; (9) Au 92.1%, Ge 7.9%; (10) U 86.6%, Rh 13.4%; (11) U 91.7%, Mo 8.3%; (12) Fe 60%, Co 40%; (13) Fe 90%, V 10%; (14) Ti 70%, Mo 30%; (15) Fe 49%, Cu 51%; (16) Ni 50%, Cu 50%; (17) Nb 95%, Ta 5%; (18) Nb 70%, Ta 30%; (19) Re 60%, Mo 40%; (20) Ti 34.3%, Zr 65.7%; (21) Co 64.5%, Ni 35.5%; (22) Mg 85%, Li 15%; (23) Mg 88%, Li 12%.

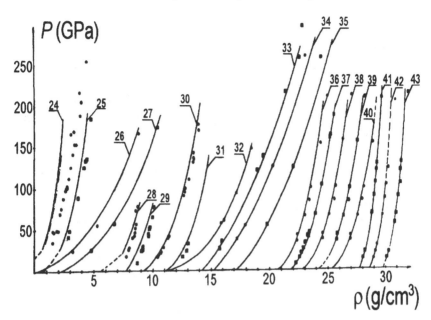

Fig. 4.2. Calculated 'additive' Hugoniots of metal alloys compared to experimental data: (24) Fe 80.2%, Si 19.8%; (25) Fe 96%, Si 4%; (26) Au 66.5%, Pb 33.5%; (27) Au 79.4%, Pb 20.6%; (28) Ti 48.6%, Ni 51.4%; (29) Ti 45%, Ni 55%; (30) Cu 61.5%, Zn 36%, Pb 2.5% (brass); (31) Pb 35%, Cd 33%, Mg 30%, Zn 2%; (32) Bi 40%, Cd 9.5%, Sn 9.5%, Pb 41%; (33) W 95%, Ni 3.5%, Fe 1.5%; (34) W 95%, Ni 3%, Cu 2%; (35) W 90%, Ni 7%, Fe 3%; (36) 25 KhGSA steel; (37) 12Kh18N10T stainless steel; (38) Steel 45; (39) 30Kh13 steel; (40) 40Kh steel; (41) AMZ; (42) AMG-6; (43) D-16 dural.

ment between the additive calculations and experimental data is quite satisfactory. From the purely formal viewpoint, the differences between calculations and measurements for the alloys Cu–W(25%), Cu–W(68%), Ti–Mo(30%), Nb–Ta(30%), Pb(35%)–Cd(33%)–Mg(30%)–Zn(2%), and brass have the same sign. But these deviations are quite small and lie within the experimental uncertainties. For example, for the Nb–Ta(30%) alloy the difference between calculated and measured densities at 4 Mbar is only 3%.

There is an apparent disagreement between calculations and experimental data in the Fe–Si and Ti–Ni alloys, which are chemical compounds. At $P = 50$ GPa this difference is about 15% for Fe–Si and about 6% for Ti–Ni. We could not remove this by using calculated Hugoniots of high-density phases of Si and Ti since the resulting deviation was still beyond the experimental uncertainty.

Analyzing these differences, we came to the following conclusions:

– firstly, the deviations of calculations from experimental data, by and large, have the same sign as and are very close to the differences between the measured zero-pressure densities of alloys and additive calculations;

– secondly, the disagreement between calculations and measurements is largest when an alloy is composed of elements with very different atomic numbers, A;

– thirdly, the deviations of calculations from experimental data are larger than the measurement uncertainty in chemical compounds, whereas alloys, which are mechanical mixtures or solid solutions, are fairly adequately described by the additive approximation.

In closing this section, note some features of the shock compression of widely used alloys of iron and aluminum. Their measurements are given in Figs. 4.3, 4.4, where they are compared to Hugoniots of almost pure aluminum (AD-1 alloy with 99.3% of aluminum) and iron (low-doped St3 steel with 99% of iron).

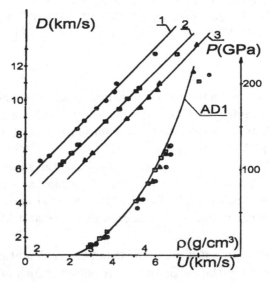

Fig. 4.3. Hugoniots of aluminum alloys: (1) D-16; (2) AMG-6 $(U+1)$; (3) AMZ $(U+2)$.

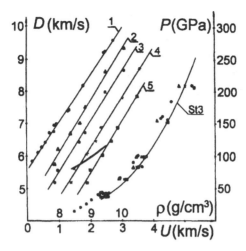

Fig. 4.4. Hugoniots of iron alloys: (1) 12Kh18N10T ($D+1$); (2) 30Kh13 ($D + 0.5$); (3) St45; (4) 25KhGSA ($U + 0.5$); (5) 40Kh ($U + 1.0$).

The experimental points only slightly deviate from the calculated Hugoniots. Since the content of impurities in Al and Fe alloys is close to that in AD-1 and St3, the assumption about the small differences between additive and experimental Hugoniots is valid.

To sum up, we may conclude that the additive calculation technique yields satisfactory results for most metallic alloys, except chemical compounds, whose additive Hugoniots should be constructed using more sophisticated methods and verified in experiments.

4.2   Compression of carbides and nitrides of metals

This group (see also Section 4.3 about hydrides) concludes our review of metal compounds (only metal phosphides, which are rarely used in technology, are not discussed in this review). These materials are also interesting in connection with the problem of compression of powders of metal carbides and nitrides for the purpose of fabricating macroscopic samples of them with mechanical parameters equal to those of crystalline materials.

Both of these types of compound are extremely hard materials. The set of studied samples [78] included AlN, Si_3N_4, TiN, ZrN, TaN (samples of tantalum and aluminum nitrides were fabricated with two different densities); metal carbides SiC, TiC, ZrC, NbC, and WC (samples of NbC also had two different densities). With the exception of porous samples, the initial densities of studied materials were close to their crystalline densities with a deviation of several percent, mostly due to a small content of impurities.

Measurements of metal nitrides are given in Figs. 4.5, 4.6. The Hugoniots are approximately parallel lines, like the Hugoniots of

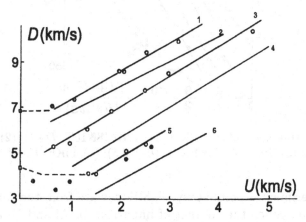

Fig. 4.5. Comparison of Hugoniots of nitrides with those of respective metals: (1) TiN; (2) Ti($D+1$); (3) ZrN ($D-1$); (4) Zr; (5) TaN ($D-1$), $\rho_0 = 14.48$ g/cm^3 (open circles), $\rho_0 = 13.74$ g/cm^3 (filled circles); (6) Ta ($D-2$).

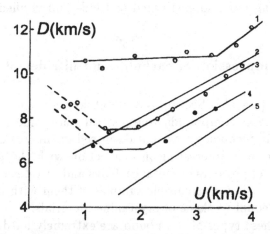

Fig. 4.6. Comparison of Hugoniots of nitrides with those of respective elements: (1) Si$_3$N$_4$ ($D+2$); (2) Al; (3) AlN ($\rho_0 = 3.23$ g/cm^3); (4) AlN ($\rho_0 = 3.00$ g/cm^3); (5) Si ($D-1$).

respective metals shown in these diagrams for comparison. On their low-density portions D = const, and it seems that these portions of the AlN and TaN Hugoniots look rather peculiar because their shock velocities drop with the particle velocity. Their shock velocities at $U \to 0$, however, tend to the speed of sound, and this indicates that measurements at low shock velocities are correct. Nonetheless, it seems that the curve shape in this region should be more complicated. The nature of this peculiar shape of the Hugoniot could be elucidated by performing experiments with samples of very different thicknesses or with continuous recording of shock parameters, such as pressure or particle velocity. But samples with the required parameters were not available for those experiments [78].

The Hugoniots of metal carbides (Figs. 4.7, 4.8) are very similar

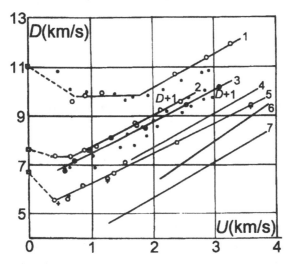

Fig. 4.7. Hugoniots of elements and their carbides: (1) SiC; (2) NbC ($D + 1$, $\rho_0 = 7.80$ g/cm^3); (3) NbC ($D + 1$, $\rho_0 = 7.29$ g/cm^3); (4) Nb; (5) WC; (6) Si; (7) W. Large circles (open and filled) correspond to the data of Russian authors, small circles correspond to the data of other authors [46].

to those in previous diagrams. The complex shapes of low-density portions may be ascribed to the effect of high compression on strengths of materials. This effect is more notable in carbides and nitrides than in elementary metals. In metal carbides the initial portions of their Hugoniots may be also affected by the transition of carbon to the diamond phase, which is highly plausible under

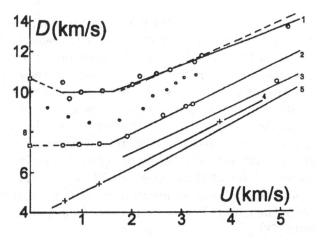

Fig. 4.8. Hugoniots of metals and their carbides: (1) TiC ($D + 1$, $\rho_0 = 4.85$ g/cm^3); (2) ZrC; (3) Ti; (4) TaC ($D - 1$); (5) Ta. Small open circles present data for TiC from Ref. [46] ($\rho_0 = 4.45$ g/cm^3).

such conditions.

In considering alloys and various compounds with different chemical bonds, crystal structures of the initial state, states of aggregation, etc. (such systems will be also discussed in the following sections), we could see that the additivity rule can be so widely applied to Hugoniots that it may be endorsed as a universal one. There are, however, natural materials to which this rule cannot be applied under any conditions. The specific volume of these materials (for example, two-component compounds) is either larger or smaller than the total specific volume of their components. This class includes AlN, SiN, and TiN. The only metal nitride whose specific volume falls between those of its components at a fixed pressure is TaN, but its additively calculated Hugoniot also disagrees with experimental data. ZrN has similar properties. Thus metal nitrides belong to the class of materials not described by the additive model.

Metal carbides are quite different. Figure 4.9 shows additively calculated Hugoniot curves with the carbon component described by the diamond Hugoniot [79] and metal characteristics determined by respective equations from Ref. [32]. You can see that additively calculated Hugoniots are in fair agreement with the entire set of experimental data, i.e., properties of all materials of this class under shock compression can be predicted using Hugoniots

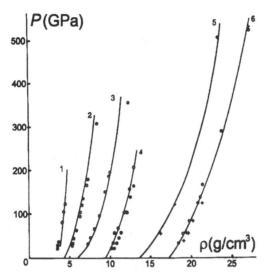

Fig. 4.9. Calculated additive Hugoniots of metal carbides: (1) SiC; (2) TiC; (3) ZrC; (4) NbC; (5) TaC; (6) WC.

of their components by calculating the 'mixed' curve.

4.3 Metal hydrides

The last class of studied metal compounds is hydrides. The research on these materials was stimulated by their widening application in various fields of technology, and interest in them was driven by the discovery of specific metal–hydrogen compounds produced under intense pulsed loads.

There is another incentive for studying metal hydrides. This is the verification of Larin's extraordinary hypothesis that planets of the Earth type, including, naturally, the Earth, are composed mostly of hydrides. In this hypothesis the high density of the Earth core is interpreted in terms of the high compressibility of metal hydrides due to the deformation of hydrogen ions, and the effectively liquid state of the outer core is related to the enhanced mobility of metal atoms in a material saturated with hydrogen at the high temperature in the Earth interior. Direct measurements of the compression of hydrides might be used to verify the underlying statement about the anomalously high compressibility of metal hydrides.

A feature of these metal–hydrogen compounds is that they can absorb a lot of hydrogen. In the process, some metals (specifically, alkali and most of alkaline-earth metals) are condensed considerably so that their low-pressure density is higher than that of pure metals. Given a large amount of accumulated data, research into these systems could clarify the role of hydrogen in metal compression and the nature of its state inside metals.

We selected for our experiments hydrides of Ca, Mg, Ti, Zr, V, and Ta. In most of them the bond between hydrogen and metal atoms is metallic, in calcium hydride it is ionic, and magnesium hydride belongs to the class intermediate between ionic and covalent compounds.

Experimental data are plotted in Fig. 4.10 and compared to measurements of respective metals. The measurements can be approximated by curves of three types: straight lines in $D - U$ coordinates (titanium and calcium hydrides), broken lines of two sections with different slopes (tantalum, magnesium, and vanadium hydrides), and two straight lines with a discontinuity at D_{cr}

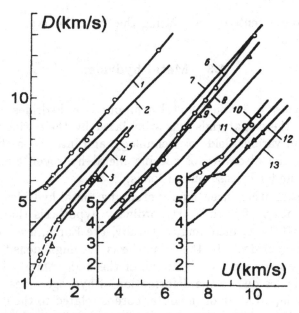

Fig. 4.10. $D - U$ curves for metal hydrides (laboratory measurements): (1) TiH_2; (2) Ti; (3) $TaH_{0.4}$ ($\rho_0 = 10.2$ g/cm^3); (4) $TaH_{0.4}$ ($\rho_0 = 12.5$ g/cm^3); (5) Ta; (6) $MgH_{1.5}$; (7) Mg; (8) $CaH_{1.9}$; (9) Ca; (10) $VH_{0.54}$; (11) V; (12) $ZrH_{1.5}$; (13) Zr.

(zirconium hydride).

The Hugoniots of hydrides in $D - U$ coordinates are quite different from those of metals, and no apparent correlations between them can be detected. It follows from the comparison of $P - \rho$ Hugoniots that the curves of compression of transition metals and their hydrides are very similar and the difference between them at maximum pressure $P \sim 200\text{--}250$ GPa is within 6%. The compressibility of a metal at a given pressure may be either higher or lower than that of its hydride.

The compressibility of hydrides of light metals (Li, Ca, Mg) is usually lower than that of pure metals. As a result of hydration, the loose structure of these metals is condensed.

The main conclusion from the experimental data is that the shock compression of hydrides does not lead to an anomalously great increase in their densities. Although the number of metals considered in this section is rather small, it seems unlikely that hydrides of other metals have radically different properties.

Thus Larin's hypothesis about the anomalously large compressibility of metal hydrides has not been confirmed, and we have to reconsider the traditional hypothesis that the iron Earth core is doped with 10–20% of lighter elements like silicon, oxygen, sulphur, etc. As concerns the role of hydrogen in hydride compression, we cannot make an unambiguous conclusion at this point, in particular, because of insufficient experimental data.

5
Minerals

Certainly, it is impossible to measure the compressibility of all natural minerals, especially as pure chemical compounds are in a minority among them, and most natural minerals are composite structures, which makes their total number much larger. Therefore we studied the most typical species of various groups of minerals and generalized our results as far as possible.

Let us recall that inorganic minerals are usually classified into four groups in terms of chemical bonds between their structural units. These are oxides, oxysalts, halides, and native elements (this group includes metals discussed in the previous section). The first three groups are subdivided into smaller subclasses, but we shall consider the compressibility of representatives of large groups only.

5.1 Oxides

5.1.1 Silicon dioxide

Of all oxide compounds whose shock compression has been measured, quartz (or, more precisely, various modifications of SiO_2) is the best studied material as concerns both the range of applied pressure, of initial states and phases, and the diversity of effects produced by shock in samples. The shock compression of α-quartz or its polycrystalline phase quartzite, coesite (one of the high-density phases of silicon dioxide), crystoballite (a less dense phase), and amorphous quartz was studied at different times for different purposes. Porous Hugoniots, double-compression Hugo-

106

niots, and expansion Hugoniots of SiO_2 have been measured. In fact, this is the only oxide whose compression was measured not only in laboratory, but also in full-scale experiments. In the latter measurements, the pressure was up to 2 TPa, i.e., nearly an order of magnitude higher than in laboratory experiments performed with such materials.

Why are researchers so fascinated with silicon dioxide? There are several reasons. One of them derives from the problem of the composition of the Earth interior since it is widely accepted among geophysicists that silicon dioxide, or its high-density modifications, is the prevailing material of the Earth mantle. Moreover, adepts of the Lodochnikov–Ramsay hypothesis about the uniformity of the Earth composition assert that the core also consists mostly of SiO_2 high-density phases. So the interest in studies of silicon dioxide was initially driven by the desire to verify these statements in shock compression experiments.

The second reason is that silicon dioxide is the most wide-spread mineral of the Earth crust, and some mountain ridges are built entirely from this material. Besides, it is a basic component of many rocks. Therefore it is reasonable to assume that the features typical of quartz also characterize, to a certain extent, rocks containing quartz. Thus the studies of their properties are closely related to those connected with silicon dioxide.

The third basic reason is that silicon dioxide is a convenient material for studies of phase transitions caused by high pressure, specifically by shock pressure. Several natural modifications of silicon dioxide are known, namely, tridymite, crystobalite, amorphous silicon dioxide, α-quartz, coesite, and stishovite. Some phase transitions between these modifications have been detected. But the question about which phase transitions actually take place under shock pressure and what is their nature has not been finally resolved at the present time. The only statement which does not cause controversy among researchers is that stishovite (or a stishovite-like liquid) is the final stage of all transformations, and the transition to this state from various initial phases takes place at a pressure of more than 20–40 GPa.

In this chapter the properties of silicon dioxide are considered in the context of the main topic of the book, namely the investigation of shock compressibility. Other problems which are beyond this topic are only formulated in this context.

5.1.2   Measurements of silicon dioxide compressibility

This research was initiated in Russia in 1960 [80], when the shock compression of α-SiO_2 was studied under a pressure of up to 30 GPa. A phase transition which was interpreted as a quartz–coesite transition was detected. A detailed investigation of quartz by Wackerle [81], which was performed in approximately the same pressure range and yielded similar results, was published at the same time. Those results were interpreted in terms of a fixed transition from α-quartz to stishovite, and this interpretation has been commonly endorsed since that time. In 1965 the upper pressure in experiments with α-quartz was 200 GPa [82], in 1970 it was 350 GPa in laboratory and 550 GPa in full-scale experiments [5]. The relative compressibility of SiO_2 with respect to an aluminum shield was measured under a pressure of up to 2 TPa. Later the compressibilities of coesite, crystoballite, fused (amorphous) quartz, the two-fold compression of porous samples, and expansion Hugoniots of SiO_2 were measured [55, 83–87].

Quartz is, in fact, the only mineral whose measured pressure ranges of laboratory and full-scale experiments partially overlap. In the latter experiments the shock propagated across a rock mass from quartzite, and the impedance-matching technique similar to that used in laboratory experiments was employed, when a quartzite sample was placed downstream of an aluminum reference sample, as well as the modified, reversed impedance-matching technique, when a sample was placed upstream of an aluminum standard sample (Fig. 2.6). In this case the shock velocity was measured over a sample thickness of 200–500 mm, whereas the sample thickness in the nuclear impedance-matching technique was commonly 80–100 mm. Since the aluminum Hugoniot plotted in the $P - U$ coordinates is very close to the quartzite Hugoniot, the sought-for states of quartzite on the $P - U$ diagram can be found in the reversed impedance-matching technique as in the common technique in which the crossing point of the $\rho_0 D_{SiO_2} U$ line with the mirror reflected aluminum Hugoniot is determined. Certainly, the wave velocities on the interface are taken. The parameters on the interface were calculated using experimental curves of shock decay (see the diagram in Fig. 5.1).

We shall not consider compression of quartz around its phase transition, which was discussed in more than ten publications by researchers from different countries and deserves special treat-

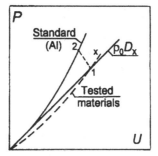

Fig. 5.1. Diagram illustrating the reversed impedance-matching technique.

ment. Here we only present the data about the position of the high-density Hugoniot branch [9, 28, 46, 81, 82, 86]. Figure 5.2 shows measurements of the compression of α-quartz (single-crystal samples) and quartzite (polycrystalline α-SiO$_2$ with the content of the main component of more than 98%).

Around the phase transition the shock velocity is a flat function $(D = \text{const} \simeq 5.75 \text{ km/s})$. This parameter was measured with electric pins placed close to the interface between shield and sample. The $D - U$ curve beyond the phase transition consists of two rectilinear portions intersecting at $D \simeq 15$ km/s $(P = 0.2$ TPa).

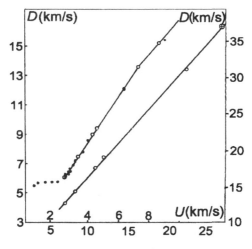

Fig. 5.2. Hugoniot of quartz (quartzite) plotted in $D - U$ coordinates: laboratory measurements (filled circles); full-scale experiments: open circles, oplus; point marked by oplus is taken from Ref. [24].

The first portion is described by the equation

$$D_1 = s_0' + \lambda_1 U \qquad (s_0' = 2.19 \text{ km/s}, \ \lambda_1 = 1.60)$$

and the second one by the equation

$$D_2 = s_0'' + \lambda_2 U \qquad (s_0'' = 4.29 \text{ km/s}, \ \lambda_2 = 1.260).$$

Note that the experimental points in this range were obtained in laboratory experiments with single-crystal α-SiO$_2$ and quartzite, and in full-scale experiments with quartzite using both conventional and reversed impedance-matching techniques. All the experimental curves coincide, which demonstrates the identity of measurements of samples with different structures and, what is more important, with different thicknesses. This proves that the shock front is the decisive factor in the kinetics of phase transitions: the new phase is generated largely on the shock front, hence this process is not controlled by the scale of the experiment. This conclusion is very important for some technologies using shock propagation across natural and artificial structures.

The experimental point at maximum pressure was obtained in full-scale measurements using the reversed impedance-matching technique. As usual, we present direct measurements plotted against the wave velocity in the shield (D_{Al}) and in quartzite (D_{SiO_2}) (curve 3 in Fig. 5.3). They are approximated by two straight lines joined at $D_{Al} = D_{SiO_2} \simeq 16$ km/s.

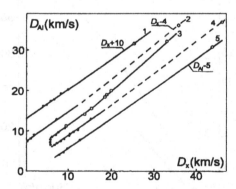

Fig. 5.3. $D - D$ Hugoniots measured using aluminum as a standard: (1) Plexiglass; (2) porous α-SiO$_2$ ($\rho_{00} = 1.75$ g/cm^3); (3) quartz (quartzite, $\rho_{00} = 2.65$ g/cm^3); (4) porous α-SiO$_2$ ($\rho_{00} = 1.35$ g/cm^3); (5) water. Laboratory measurements (dots), full-scale experiments (open circles).

In deriving other parameters of the compression of silicon dioxide from the measured shock velocities, we used a version of the impedance-matching method. The starting aluminum parameters were the particle velocity and pressure derived using the equation

$$D_{Al} = 5.90 + 1.19U \qquad (\rho_0 = 2.71 \text{ g/cm}^3),$$

which was given above. The $P - U$ curves were plotted using the mirror reflected aluminum Hugoniot.

The data on quartz compression are plotted in the $P - \sigma$ coordinates in Fig. 5.4. Beyond the quartz–stishovite phase transition,

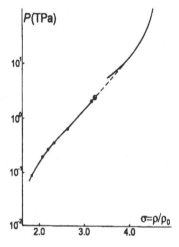

Fig. 5.4. $P - \sigma$ Hugoniot for silica. Extrapolation to the region in which the Thomas–Fermi model is valid. Laboratory measurements (filled circles) , relative measurements of full-scale experiments (oplus and open circle).

the experimental curve is monotonic up to a pressure of $P = 2.8$ TPa, which indicates that the phase state does not change. A calculated extrapolation of experimental curves into the high-pressure region is also shown in the diagram. The calculated SiO_2 Hugoniot was obtained by adding specific volumes of silicon and oxygen with weights corresponding to their stoichiometric coefficients at fixed pressures and temperatures. The individual Hugoniots of silicon and oxygen were calculated by the Thomas–Fermi model.

The fact that the interpolation curve and the calculated Hugoniot gradually approach each other in the high-density limit is in favor of the assumption about the monotonic shape of the quartz compression curve in the high-pressure limit. At the same time, the coincidence of the calculated and experimental curves justifies the application of the additive technique to constructing high-pressure Hugoniots of compound materials using the parameters of their constituent elements. This observation confirms calculations of Hugoniots of many materials, rocks, and other composite structures.

The smooth shape of the quartzite Hugoniot leads us to the conclusion that the experiments on quartzite compression (at an initial density $\rho_0 = 2.65$ g/cm^3) contradict the Lodochnikov–Ramsay hypothesis about the dielectric-to-metal transition in silicate systems, whose basic component is SiO_2, under the pressure at the Earth mantle–core interface ($P = 140$ GPa).

5.1.3   Dynamic compression of porous silicon dioxide

The aim of the experiments discussed in this section was to investigate the nature of phase transitions in silicon dioxide and to obtain information about the thermal component in its equation of state.

The conclusion that the phase transition in silicon dioxide is between α-SiO_2 and stishovite [81, 82] was based on the interpolation of the steep 'stishovite' branch of the Hugoniot to zero pressure, which yielded a low-pressure density quite close to that of stishovite, $\rho_0 = 4.28$ g/cm^3. (The controversy about this transition led to an alternative interpretation [80].) This conclusion [81] could be justified by measurements of the shock compression of quartz of various initial densities if they yielded a fan of Hugoniots originating from the point of the stishovite low-pressure density, ρ_0, provided that the α-SiO_2–stishovite transition really occurred on these curves. In reality, the situation proved to be, as often happens, more complicated, and the nature of the phase transition in silicon dioxide under shock has not been finally established to this day. Nonetheless, the issue of the transition from α-quartz to stishovite (or stishovite-like liquid) has been settled in this research. Before discussing the results, let us make the following remarks.

The first one is about the starting materials. As was mentioned above, there are several natural modifications of silicon dioxide. It was of interest to find out at least the positions of Hugoniots of some of these modifications if their initial densities were equal, i.e., whether their positions depended on the initial structure.

The second remark concerns the range of initial silicon dioxide densities. It turned out, for example, that a material with a density of less than $\rho_{00} = 1$ g/cm^3 could not be fabricated from α-quartz powder with a grain dimension of less than 100 μm. To overcome this difficulty, we had to use a finer powder of nonaqueous silicic acid (SiO$_2$), from which samples with an initial density $\rho_{00} = (0.5$–$0.8)$ g/cm^3 were manufactured. Samples of lower density $\rho_{00} = (0.2$–$0.5)$ g/cm^3 were fabricated from Aerosil (SiO$_2$) powder with submicron dimension.

The final remark is about preliminary experiments undertaken to study the effects of various factors on measured parameters. These are (they were mentioned in the section about porous metals) the humidity of samples, grain dimensions, presence of air in samples, additional spread of the shock front, and impurities. The preliminary experiments were performed with prebaked samples; with materials in which the grain dimension varied between 0.01 μm and 0.3 mm; with two samples of equal initial density, but with different impurity content (samples with $\rho_{00} = 0.55$, 1.15, and 1.75 g/cm^3 were used); with samples from which air was evacuated down to a pressure of 0.3–0.5 mbar (materials manufactured from silicic acid with $\rho_{00} = 0.55$ g/cm^3 and quartz powder, $\rho_{00} = 1.75$ g/cm^3, the shock pressure was 14 and 30 GPa, respectively).

After direct laboratory measurements [55, 84] we came to the conclusion that the results of preliminary experiments coincided within their measurement uncertainty with those of the main series of experiments, therefore all the data were accumulated in a summarized set of measurements. No additional spread of the shock front (as compared to experiments with solid metals at equal shock amplitudes) was detected [55].

All experimental data [55, 83, 84] are plotted in Fig. 5.5. Beside Hugoniots of porous silicon dioxide, the diagram also contains data about coesite ($\rho_0 = 2.92$ g/cm^3). All in all, thirteen Hugoniots of silicon dioxide with densities ranging between 0.2 g/cm^3 and 2.92 g/cm^3 (to say nothing of wide variations in the compositions of samples with equal initial densities) were recorded. Note that the Hugoniots can be classified into two groups. The first group includes those with an initial density varying between 1.15 and 2.92 g/cm^3. Their common feature is a discontinuity on the $D - U$

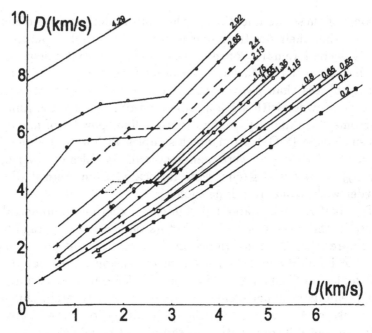

Fig. 5.5. Hugoniots for silica. Laboratory measurements.

curve, which is usually related to a first-order phase transition. The second group includes Hugoniots with $\rho_{00} \leq 0.8\,\mathrm{g/cm^3}$, which have no such discontinuities.

Generally speaking, the layout of Hugoniots in Fig. 5.5 is similar to that of $D - U$ curves for porous metals: here we can also see a fan of diverging curves with a variable slope and a convergence of Hugoniots to the boundary curve $D = U$. The only obvious difference is that the curves for porous SiO_2 do not have a clearly defined point s'_0, from which all porous Hugoniots should originate. It is difficult to assert whether there is an underlying physical cause of this or if it is due to a large experimental error, which may take place at low pressure.

Measurements of the shock compressibility of highly porous silicon dioxide ($\rho_0 \simeq 0.2\,\mathrm{g/cm^3}$) by several authors [55, 88, 89] are compared in Fig. 5.6. The data from the first two publications correlate well to each other. Their absolute measurements are very close, and the $D - U$ curves are parallel. The experimental data on SiO_2 with initial density of $0.13\,\mathrm{g/cm^3}$ are approximated by a single straight line [88] and the Hugoniot with the initial

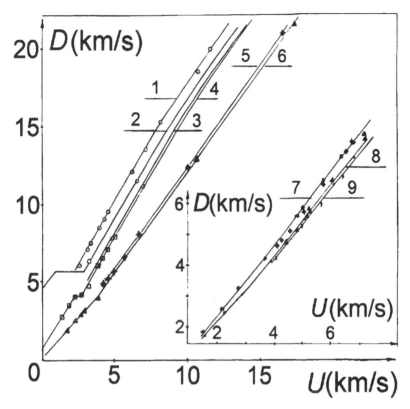

Fig. 5.6. Relative measurements of highly porous silica: ρ_{00} (g/cm^3) = 2.65 (1), 2.2 (2), 1.75 (3), 1.35 (4), 0.34 (5), 0.195 (6), 0.4 (7), 0.2 (8), 0.13 (9). Different symbols present data from Refs. [55, 88, 89].

density $\rho_{00} = 0.2$ g/cm^3 is quite close. A small difference between slopes of lines approximating laboratory and full-scale measurements [55, 88, 89] is not unusual since at very high shock velocities the slopes of all Hugoniots tend to a common limit $D' \simeq 1.25$. The possible uncertainty in particle velocities in SiO$_2$ measured at maximal kinematic parameters [89] may be caused by the error due to the difference between the mirror reflected Hugoniot and expansion isentrope. In this case, the correction is positive (the particle velocity U should be increased), but its specific value depends on the form of the selected equation of state.

Now let us consider the $P - \rho$ curve of SiO$_2$ (Fig. 5.7). The Hugoniots plotted in this diagram demonstrate a complicated and not easily interpretable pattern of compression curves for different

116   *Minerals*

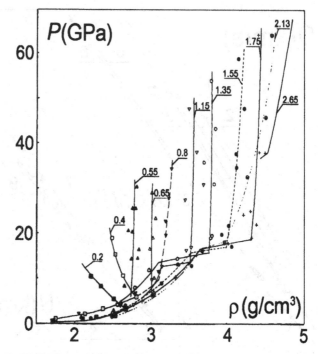

Fig. 5.7. $P - \rho$ curves for silica. Laboratory measurements.

initial states of silicon dioxide. The curves may be divided somewhat arbitrarily into three groups. The first group includes Hugoniots with an initial density $\rho_{00} > 1.55$ g/cm^3. The straight lines
extrapolating their steep high-pressure portions to $P = 0$ cross the
density axis near the point with $\rho_0 \simeq 4.1$ g/cm^3, which is about
the low-pressure density of stishovite. In this group the Hugoniot corresponding to the low-pressure density $\rho_{00} = 2.13$ g/cm^3
crosses the curves with $\rho_{00} = 1.55$ and 1.75 g/cm^3, which is not
admissible in the model of state in thermodynamic equilibrium.
Maybe this indicates that stishovite branches of Hugoniots with
$\rho_{00} > 2.1$ g/cm^3 slightly above the phase transition correspond
to nonequilibrium states in which a small amount of the initial
phase is mixed with the stishovite phase, and the entire material
is transformed to the high-density phase only at a higher pressure
[85]. This may also account for the deviation of lines extrapolating the stishovite branches of the Hugoniots from the initial
stishovite state. Note that the slopes of three Hugoniots with
$\rho_{00} > 2.1$ g/cm^3, $(dP/d\rho)_H$, are close to each other and different

from the other two Hugoniots with $\rho_{00} = 1.55$ and 1.75 g/cm^3. The compression Hugoniots of SiO$_2$ with $\rho_{00} < 0.8$ g/cm^3 belong to the second group, and it is natural to classify them as the coesite type. In fact, lines extrapolating high-pressure branches of these Hugoniots to zero pressure yield initial densities close to that of coesite ($\rho_{00} = 2.98$ g/cm^3). Another distinction of these curves from those of other groups is that in $D - U$ coordinates they have no portions with $D = $ const.

The third group includes three Hugoniots with $\rho_{00} = 1.15$ and 1.35 g/cm^3. Their extrapolations yield densities ranging between those of stishovite and coesite.

The interpretation of high-density states of experimental silicon dioxide Hugoniots proposed in this section seems more plausible than any other.

To conclude this section, let us consider the results of full-scale experiments with porous samples of α-SiO$_2$ with $\rho_{00} = 1.15$ and 1.35 g/cm^3. These data [9] were obtained in 1970. An aluminum shield across which the shock propagated towards the samples had a thickness of about 100 mm, and that of the samples was about 80 mm. The device was located at about 6 m from the explosion center, i.e., the distance was fairly large. As a result, the shock attenuation over the sample thickness was relatively small. The drop in shock velocities was calculated to be less than 1% of the average shock velocity. The shock velocities calculated on the aluminum–sample interface and parameters of the porous quartz dynamic compression are listed in Table 5.1 and plotted in Figs. 5.8, 5.9. In the former diagram the results are compared to laboratory [83, 84, 89] and full-scale measurements [5, 9, 17, 24, 89]. Almost all the data form a self-consistent pattern, only the point for

Table 5.1.

Parameters in the shield		Studied material	Shock parameters in studied material			
D	U	ρ_0	D	U	P	ρ
km/s	km/s	g/cm^3	km/s	km/s	GPa	g/cm^3
36.40	25.63	H$_2$O, 1.00	45.95	32.54	1430	3.85
22.10	13.61	Plex., 1.18	25.19	16.96	504	3.61
36.70	25.88	SiO$_2$, 1.75	39.73	29.08	2022	6.53
37.59	26.63	SiO$_2$, 1.35	42.17	31.89	1815	5.54

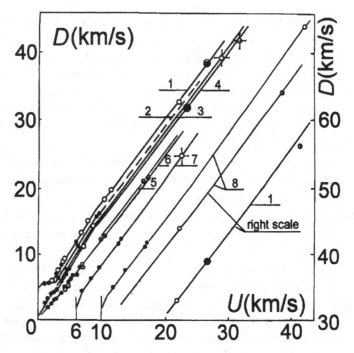

Fig. 5.8. Hugoniots of lighter materials: (1–6) silica with densities
$\rho_0 = 2.65, 2.20, 1.75, 1.35, 0.34$ and 0.195 g/cm^3; (7) Plexiglass; (8) water. Laboratory data: dots; data of full-scale experiments: open circles, crossed circles, filled circles, and half-filled circles.

fused quartz [24] is controversial, whereas parameters of the point taken from Ref. [17] were derived by measuring the diagram given in the paper and it is only an approximate indication of the $D-U$ Hugoniot position.

To sum up, the Hugoniots of silicon dioxide with different initial densities form a pattern of approximately parallel lines with a variable slope. At velocities less than $D = 14$–16 km/s their slope $D'_U = 1.6$, and at higher velocities 1.2–1.3. Note that the change in the slope depends on the initial density of silicon dioxide.

The uncertainty of these measurements was estimated to be within 2% of the average velocities (it is shown in Fig. 5.8 by vertical arrows near experimental curves), i.e., the experimental error was rather large. Nonetheless, the good coincidence of various measurements indicates that the most probable values of measured parameters were determined fairly accurately.

Figure 5.9 shows the curves plotted in $P-\rho$ coordinates. These

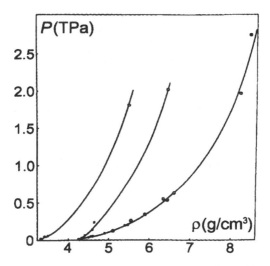

Fig. 5.9. $P - \rho$ curves of silica. Laboratory data (dots), full-scale measurements (open circles).

are the usual parabolic curves with rapidly rising slopes.

By comparing Hugoniots with $\rho_0 = 2.65$ and 1.75 g/cm^3 (both of them are stishovite-like curves), we can estimate the average Grüneisen parameter, $\overline{\Gamma} = \Delta P_T/\rho \Delta E_T$ (the index T indicates the thermal components of the pressure and energy). We assume that both compared parameters, P and E, refer to the liquid phase of silicon dioxide. We obtain $\overline{\Gamma} = 0.6 \pm 0.1$, which is a reasonable value for these states. Note that similar parameters are obtained when the Hugoniot with $\rho_{00} = 1.35$ g/cm^3 (let us recall that it does not belong to the stishovite group) is compared with the $\rho_{00} = 2.65$ g/cm^3 Hugoniot. A smaller parameter $\overline{\Gamma} \simeq 0.5$ is derived from the comparison of data [89] with $\rho_{00} = 0.32$ and 0.19 g/cm^3 measured at considerably higher temperatures.

5.1.4   Double compression of quartzite

One aim of our research was to discover a silicon dioxide phase of a density higher than that of stishovite. The discovery of such a phase would be favorable for modelling the Earth interior on the basis of the hypothesis that its composition is uniform with a considerable content of high-density SiO$_2$ phase.

We could not measure stishovite compressibility directly because samples of sufficient dimensions were not available.* Results of experiments with α-SiO_2 and other modifications were not encouraging, as one could see in the data of previous sections. No features that could be connected to a transition to a denser phase were detected on the branch of the polycrystalline quartz Hugoniot related to the stishovite phase under a pressure of up to 2.8 TPa.

One possible reason may be a high increase in temperature caused by shock compression of quartz (since its 'porosity' is higher than that of stishovite), which prevents formation of a superdense phase. The sample heating can be reduced if we use the double compression technique [28, 91], in which a sample of quartz is transferred to the stishovite branch by the first shock and then compressed to a higher density by the second shock.

In experiments of this kind, the first shock passed across a tested sample and the second shock was a reflection from a harder standard material (usually metal) placed downstream of the quartz sample. The shock velocities in both materials were measured. Given their Hugoniots, parameters of doubly compressed quartz can be calculated easily: $P_2 = P_1 + \rho_1 D_{12}(U_1 - U_2)$, $\rho_2 = \rho_1 D_{12}/[D_{12} - (U_1 - U_2)]$, where D_{12} is the velocity of the reflected shock with respect to that of the incident shock wave. A phase transition on a double compression Hugoniot is described by similar equations. In this case P_2, ρ_2, and U_2 are parameters of a cusp on the Hugoniot, and U_2 and D_{12} describe the region of phase transition and correspond to the parameters U_3 and D_{23}. A detailed discussion of the measurements and experimental accuracy (ten to thirty measurements in five to fifteen separate experiments were performed to minimize the experimental error), kinetics of the processes and geophysical aspects can be found elsewhere [85, 87].

This review presents the final results (Figs. 5.10, 5.11). In the former diagram the data are plotted in the coordinates $D_{12} - \Delta U_{12}$, where $\Delta U_{12} = U_1 - U_2$. The initial values of D_{12} correspond to the speed of sound on the first Hugoniot. The curve of D_{12} versus U_{12} is nonmonotonic and has a maximum at $\Delta U = 1$km/s[†].

---

* Recently [90] the first measurements of stishovite compression have been published, but these data have very large uncertainties.
† For comparison the double compression Hugoniot of Al (see Fig. 5.10) does not display singularities similar to those of SiO_2.

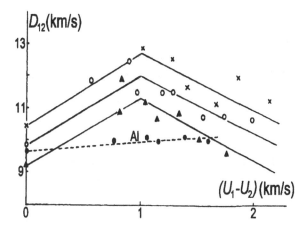

Fig. 5.10. Hugoniots for aluminum and silica measured in double compression experiments. Filled circles represent data on aluminum.

As a result, the compression curves in the $P - V$ coordinates have cusps (Fig. 5.11), which can be interpreted in terms of the phase transition from stishovite to a new phase. These shapes of the Hugoniots are similar to curves with phase transitions under shock. The first steeper portion corresponds to the stishovite state, apparently with an impurity of quartz, and the flatter portion corresponds to a mixture of stishovite with a denser phase,

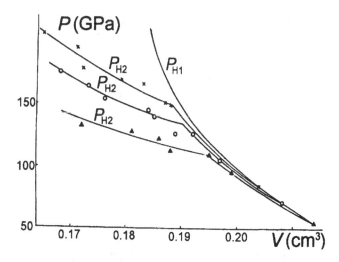

Fig. 5.11. $P - V$ diagram of double compression curves of quartz. P_{H1} and P_{H2} Hugoniots correspond to single and double compression.

probably with the fluorite structure. A steep portion corresponding to the most dense phase was not recorded in these experiments.
It seems that fundamentally new techniques generating the second
shock independent of the primary pulse are needed to detect the
new phase.

Given the abundance of experimental data [87], there is no reason to question the results of this research. Nonetheless, direct
measurements of stishovite compressibility are of paramount importance because they would make the interpretation of data more
straightforward and close the discussion about an SiO_2 phase with
a density higher than that of stishovite.

5.1.5   Compression of rutile, cassiterite, magnetite, perovskite, and periclase

Figure 5.12 shows a set of experimental data about several oxides
[28, 46, 92, 93]. The diagram contains Hugoniots with crystalline
density at zero pressure. Compression of porous samples of rutile
(TiO_2) and periclase (MgO) was also studied [94, 95].

All the Hugoniots are discontinuous which indicates phase tran-

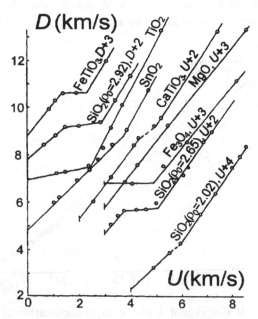

Fig. 5.12.   Hugoniots of oxides, laboratory data.

sitions in these materials caused by shock or transitions from elastic to plastic deformation leading to changes in their densities in the yield region.

It was stated in the previous section that all the modifications of silicon dioxide with a density higher than 1.55 g/cm^3 are transformed at a certain shock pressure to stishovite, which has a structure similar to that of rutile with a coordination number of six. What will happen to rutile under these conditions? Its $P - \rho$ diagram, which contains the 'normal' Hugoniot and also the compression curves of porous samples, is given in Fig. 5.13. The shape of the high-pressure branches indicates that there is a high-pressure phase of rutile with $\rho_0 \simeq 6.0$ g/cm^3.

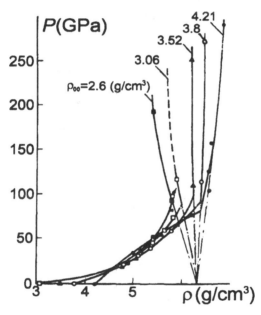

Fig. 5.13. $P - \rho$ diagram of rutile. Numbers on the curves are initial densities.

A similar phenomenon is observed in cassiterite (SnO_2) since the extrapolation line of the high-pressure branch of its Hugoniot yields $\rho_0 \simeq 11$ g/cm^3. Therefore we may assume that solids with crystal lattices of the rutile type undergo a phase transition to a denser phase. According to the Goldschmidt rule, its structure is more densely packed and has a higher coordination number of eight, like fluorite (CaF_2). Whether a similar transition takes place in stishovite, whose structure is similar to that of rutile, was discussed in the previous section. The compression curves of TiO_2 and SnO_2 indicate that such a transition is possible.

5.2 Oxysalts

This group includes:
- carbonates: calcite $CaCO_3$, magnesite $MgCO_3$, and dolomite $CaMg(CO_3)_2$ [96];
- sulphates: barite $BaSO_4$, anhydrite $CaSO_4$, and gypsum $CaSO_4 \cdot 2H_2O$ [92];
- silicates: topaz $Al_2SiO_4[FOH]_2$, enstatite $MgSiO_3$, microcline $K[AlSi_3O_8]$, beryl $Be_3Al_2[Si_6O_{18}]$, hedenbergite $CaFeSi_2O_6$, tremolite $Ca_2Mg_5[Si_4O_{11}]_2[OH]_2$, wollastonite $Ca_3[Si_3O_9]$, spodumene $LiAl[Si_2O_6]$, and nepheline $Na[AlSiO_4]$ [92, 93, 97].

The measurements are given in Figs. 5.14, 5.15. Most samples were natural and relatively pure. The content of impurities was

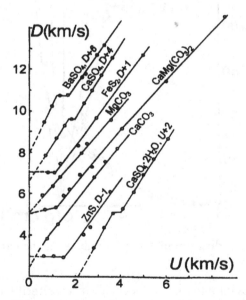

Fig. 5.14.   Hugoniots for carbonates, sulphates, and sulphides.

within 3% (except nepheline and microcline, in which their contents were 10 and 7%, respectively).

Hugoniots of materials of different groups are different, but within one group the shapes and slopes of their high-pressure branches are very close. The natural explanation of this similar-

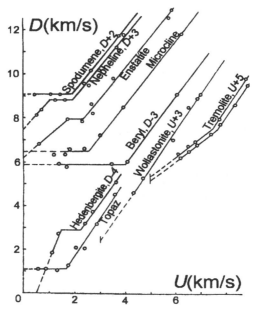

Fig. 5.15. Hugoniots for silicates.

ity is the likeness of the structures of minerals within one group. For example, silicate minerals include the same structural units, namely silicon oxide tetrahedrons, from which simple or complex groups (radicals) are built. The slopes of high-pressure portions of their Hugoniots are most probably controlled by the compressibility of silicon oxide cells. The diversity in the low-pressure slopes of silicate Hugoniots can be ascribed to the difference in the packing density of tetrahedrons.

5.3 Halides and sulphides

We have studied at different times and with different aims the following materials of this group: Na_3AlF_2, CaF_2, BaF_2, NaCl, CsI, LiF, KBr, NaI, KCl, MgF_2, CsBr, and the sulphides ZnS and FeS_2. The Hugoniots of halides [92, 98–100] are given in Fig. 5.16. The most interesting of them are those which have some singularities. From this viewpoint, a typical example is the phase transition at $P = 2$ GPa in KCl and KBr detected by all three techniques used in our experiment, namely, the shorting-pin technique, in which the shock velocity is measured, electromagnetic technique measur-

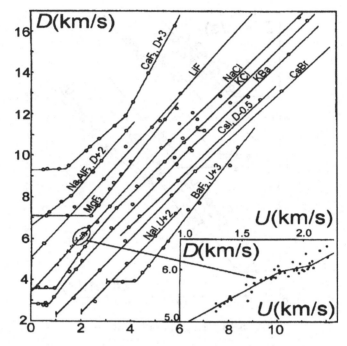

Fig. 5.16. $D - U$ diagrams for halides.

ing the particle velocity, and that using a manganin transducer in which the electric resistance of a shock compressed wire is measured. The Hugoniots of these compounds are clearly divided into three portions: on the first portion the initial phase with a body-centered cubic cell similar to that of NaCl and a coordination number of six is compressed, the second portion where $D = $ const describes a mixture of the low and high-density phases and the shock leading edge has two steps in this range of parameters, and on the third portion the high-density phase is compressed. Like CsCl, the dense phase of KCl and KBr has a crystal structure with a coordination number of eight.

A second jump of the particle velocity was detected on the Hugoniots of KCl at 150–180 GPa and KBr at 180–190 GPa [100] and this jump was ascribed to a phase transition in the liquid state.

The shock compression of rock salt should be discussed in detail. In static experiments, a phase transition in this material was detected by X-ray diffraction under a pressure of 20 GPa [101, 102]. But in dynamic experiments performed both in Russia and the

USA [46, 47, 103] no convincing evidence of a phase transition was found on the Hugoniots under pressures of up to 170 GPa, although in some experiments [104–106] small deviations of measurements from monotonic curves were detected in this range, and which be interpreted in terms of a phase transition.

Convincing proof in favor of a phase transition was obtained by analyzing records of shock wave velocity in rock salt due to a nuclear explosion [107]. Curves derived from the data of several experiments are shown in Fig. 5.17. Having selected the parameters D and t/r, where t is the time and r is the distance from the source, we can present all the data in a single curve of the shock velocity versus the reduced coordinate. Its shape is peculiar: at

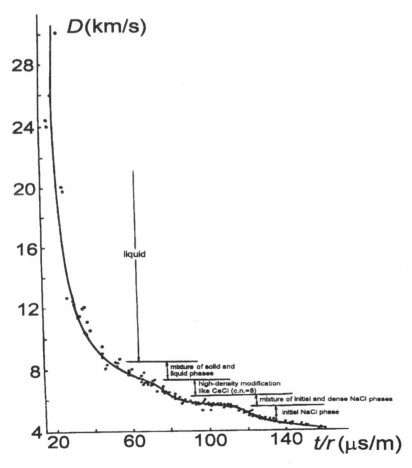

Fig. 5.17. $D(t/r)$ curve for rock salt.

a large reduced distance the material is in the low-density phase,
in the intermediate range a transition from the low-density to the
high-density phase takes place, then follows the region of the liquid
state. Given this interpretation of the data, we had to reconsider
the $D - U$ Hugoniot of rock salt. Additional laboratory experiments were undertaken, specifically in the shock velocity range of
the phase transition detected in underground experiments. These
measurements are plotted in Fig. 5.18.

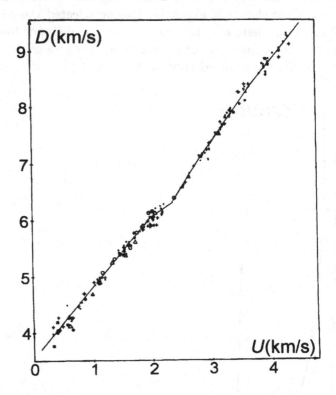

Fig. 5.18. Polymorphic transition in NaCl. Data of Russian researchers: open circles and triangles; data from laboratories in other
countries: downpluses, dots and pluses.

A $D - U$ curve of complex shape was recorded in this velocity
range. Although the particle velocity jump at $D'_U \geq 5.8$ km/s
is rather small (therefore it was not detected in first experiments), the changes in the curve slope are fairly sharp: on the
first Hugoniot portion $D'_U = 1.35$, in the range of the mixed
phase $D'_U = 0.53$, on the second portion $D'_U = 1.57$, and at

$D > 8.0$ km/s $D'_U \simeq 1.2$. These changes are indications in favor of phase transitions taking place in laboratory experiments.

At high pressures $(P > 170 \text{ GPa})$ a notable change in experimental parameters was detected [100] and it was interpreted as a phase transformation in the liquid state, as in KCl and KBr. The changes in the parameters on the Hugoniot were fairly large: the density jump was 9%.

Note that since the publication of these data [100] no attempt has been made to verify this conclusion in laboratory experiments, although some doubt in its correctness remained. This doubt derives from the fact that similar features on the aluminum Hugoniot [33, 108] were detected using the same experimental facility, but they were not found in later experiments [15, 18, 45].

Furthermore, we should stress that the curves of shock decay shown in Fig. 5.17 are monotonic, indicating the absence of phase transitions in NaCl under these conditions. This is earnest evidence that the data of Ref. [100] are inaccurate at maximum shock velocities and that the interpretation of KCl and KBr measurements in the liquid state in terms of a phase transition is erroneous. Note that measurements of NaCl parameters in underground tests present a rare example of full-scale experiments being used to verify laboratory data.

Now let us briefly discuss fluorite compression. Its Hugoniot consists of three portions: the first one is horizontal, the second is relatively flat with a slope $D'_U = 1.2$, and the third one is steep with $D'_U = 2.2$, which is, probably, the high-pressure branch. On the other hand, the Hugoniots of other fluorite structures–BaF$_2$ and MgF$_2$–have different shapes. Probably, this difference is due to different critical pressure of a transition to a dense phase with a low compressibility.

If we estimate the density of the metastable fluorite phase by extrapolating its high-density Hugoniot branch to zero pressure, we obtain for the new phase $\rho_0 \approx 1.7\rho_0$. We can account for this large increase in the density in terms of a transition from the fluorite structure with a coordination number of eight to the MgCu$_2$ structure with a coordination number of twelve or another densely packed phase. With this approach to the phase diagram, we can reconsider the hypothesis about the uniform chemical composition of our planet.

The Hugoniots of two sulphides ZnS and FeS$_2$ are plotted in Fig. 5.14.

6

Rocks

The first short note about an investigation of rock compression was published in 1957 [109]. It reported single measurements of the compressibilities of two rocks–gabbro and dunite.

In our first publication about deep rocks we discussed mostly olivine and enstatite [97]. The pressure applied to rocks in those experiments remains a record high to this day, and it was in fact the first study in which the compressibility of basic and ultrabasic rocks was systematically investigated with due regard to geophysical problems. Most of the Hugoniots given in that paper (Fig. 6.1) were straight lines of approximately equal slopes with various discontinuities on $D - U$ curves caused by phase transitions. Those

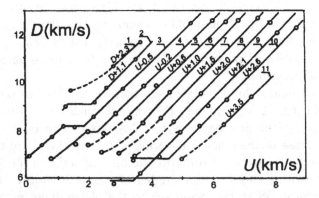

Fig. 6.1. Hugoniots of deep (basic and ultrabasic) rocks: (1) olivine diallagite; (2) ore olivinite; (3) olivinite-I; (4) enstatite rock; (5) olivinite-II; (6) feldspar peridotite; (7) dunite-II; (8) enstatite gabbro with olivine; (9) dolerite (traprock); (10) olivine diabase; (11) dunite-I.

measurements of enstatite and olivine rocks were the first experimental data used to determine the content of iron in the Earth lower mantle.

The measurements published by researchers from other countries [46, 47] also contain a lot of information, but their experiments were performed at much lower pressures.

The data about crustal rocks are rather scarce. The most notable publications on this topic in terms of the amount of information are Refs. [12, 28], which summarized measurements of more than one hundred magmatic and sedimentary rocks. Let us recall that the first class includes large groups of granites, prophyries, gabbros, syenites, tuffs, etc., the second class includes sandstones, siltstones, clays, slates, limestones, and dolomites.

Measurements of magmatic rocks are given in Fig. 6.2. The curves averaged over different rocks and some experimental points are plotted at the bottom of the diagram. The points define a fairly narrow strip, whose width characterizes the maximum spread of parameters of various rocks. The curve can be divided into three portions: the flat portion with a slowly changing shock velocity, the second steeper portion with a slope $D'_U \simeq 1.5$–1.6,

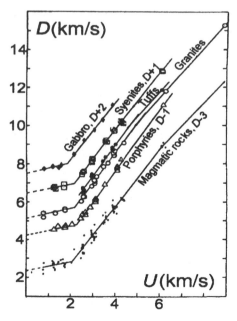

Fig. 6.2. Hugoniots of magmatic rocks.

and the portion with a slope $D'_U = 1.2$–1.3. The transition to
the slope $D'_U = 1.25$ takes place in all magmatic rocks at approximately equal shock velocities $D_U = 11$–12 km/s. The slopes of
the second and third portions of Hugoniots are about the same as
those of curves for silicate minerals and quartz, which is the basic
oxide in the rocks. As concerns the first portion, it is possible that
parameters of individual minerals are averaged under pressure because of the complex composition and structure of rocks, unlike
quartz and silicate minerals, whose Hugoniots are different. As a
result, the shock parameters of rocks are similar and described by
the flat portion of the $D - U$ curve, which is different from those
of individual minerals.

Figure 6.3 shows Hugoniots of sedimentary rocks. The similarity observed in the case of magmatic rocks cannot be seen
in this diagram. With the exception of slates and sandstones,
which are like magmatic rocks, they have radically different $D - U$
curves. The Hugoniots of calcite (limestone, $CaCO_3$) and magnesite ($MgCO_3$) in the studied velocity range are straight lines
and have kinks in their $D - U$ curves (the transition from calcite
to aragonite) at pressures considerably lower than in the case of

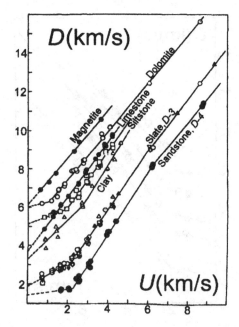

Fig. 6.3. Hugoniots of sedimentary rocks.

magmatic rocks. The Hugoniot of dolomite $(CaMg(CO_3)_2)$ has a narrow initial flat portion and then it is a straight line up to a velocity of 16 km/s. The Hugoniots of clays, whose low-density portions have slopes much higher than those of magmatic rocks, are also different.

Now let us discuss the reproducibility of Hugoniots of crust rocks. In Section 5.2 we could see that the quartzite Hugoniot was independent of the sample thickness. But the properties of SiO_2 cannot be extended to other types of rocks, especially to those which do not contain any quartz whatsoever. The effect of dimension on their Hugoniots has not been studied.

We investigated the dimensional effect in two surface rocks–slate and dolomite [12]. In the experiment we used the impedance matching technique with aluminum used as a standard material. The results and laboratory measurements are plotted in Fig. 6.4, and the coincidence of data is obvious. The independence of the three rocks–dolomite, slate, and quartzite–from the sample dimension is a strong argument that the dimension is also not essential in other rocks. As was noted above, this demonstrates that the shock front plays a dominant role as a generator of a new phase under a shock load.

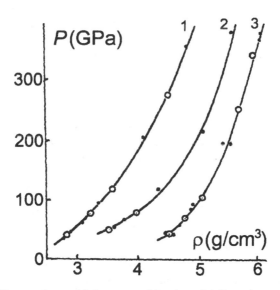

Fig. 6.4. Comparison of laboratory (dots) and full-scale measurements (open circles) of slate (1), dolomite (2), and quartzite (3).

6.1   Calculation of rock parameters

Given a lot of experimental data about the compression of various rocks and minerals (we have data on two hundred materials), we can plot generalized $D - U$ curves using common features of experimental Hugoniots, namely:

- linear $D - U$ Hugoniots of high-density phases in certain ranges of wave velocities;

- approximately equal slopes, D'_U, for many rocks and minerals;

- phase transitions in most rocks and minerals.

Using the accumulated data about rocks and minerals, we investigated the correlation between oxide composition and shapes of Hugoniots [110]. Our approach was similar to Birch's well known concept which stipulated that under high pressure the mineral structure can be adequately described as a mixture of silicon and metal oxides.

This concept, based on the rule of additivity, has been previously tested many times [82, 111]. A comprehensive test of this approach was undertaken by Al'tshuler and Sharipdzhanov [111] but their technique includes a long and labor-consuming calculation on a computer, which is not acceptable in some cases.

We used a simple technique [110] of regression analysis of correlations between experimental data about many rocks and minerals, including carbonate, silicate, and oxide rock components, with a view to deriving simple analytical equations

$$s_0 = s_0(Z, \rho_0), \qquad a_0 = a_0(Z, \rho_0),$$

which can be used to calculate Hugoniots if the equation $D = s_0(Z, \rho_0) + a(Z, \rho_0)U$ adequately describes experimental data.

The functions s_0 and a_0 were expressed by equations of multiple regression:

$$s_0 = a_0 + a\rho_0 + \sum a_i Z_i, \qquad a = b_0 + b\rho_0 + \sum b_i Z_i,$$

where Z_i is the relative content of the i-th component in the material, a, a_0, a_i, b_0, b, and b_i are coefficients of the regression equations. The numerical coefficients are listed in Ref. [110]. Calculated curves for some minerals are compared to experimental points in Fig. 6.5, and the agreement is satisfactory. Curves of minerals are approximated with a similar accuracy.

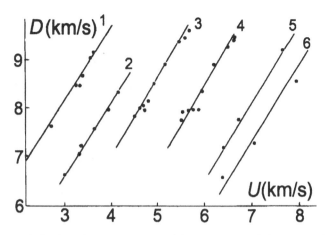

Fig. 6.5. Calculated Hugoniots of (1) magnesite, (2) calcite $(U + 1)$, (3) SiO_2 $(U + 2)$, (4) H_2O $(U + 3)$, (5) MgO $(U + 4)$, (6) Fe_2O_3 $(U + 4)$.

The analysis of calculated curves [110] demonstrates that the calculation error at a pressure of up to about 200 GPa is approximately equal to the experimental uncertainty. Thus the chemical composition of a rock or mineral is sufficient to plot its Hugoniot.

To conclude this section, let us discuss the algorithm for calculating Hugoniots of rocks in the range of parameters where theoretical models based on the Thomas–Fermi approximation are valid. In order to estimate Hugoniots (a scheme of exact calculations for such multicomponent systems as rocks is not available), we use additive equations:

$$V_r = \sum_i \alpha_i V_i(P, T), \qquad E_r = \sum_i \alpha_i E_i(P, T), \qquad i = 1, 2, \ldots$$

where V_i and E_i are the specific volume and inherent energy of rock components, and α_i is the relative weight content. We presume that all the components contained in the rock in the form of small particles are in thermal equilibrium, i.e., their pressures and temperatures are equal. At the same time, each component has its own equation of state, i.e., its particles are sufficiently large.

As was stated above, this approach proved to be correct in the case of quartzite since measurements of its compression at a pressure of about 2 TPa could be well fitted to the short-range interpolation between earlier experimental data and high-pressure calculations. Unfortunately, this was the only practical proof of the method. Nonetheless, interpolation curves of a longer range between experimental points and high-density calculations using the additive equations are always good. Figure 6.6 shows as an illustration experimental and calculated Hugoniots of granite, dolomite, and slate. The interpolating curves smoothly connect experimental points with calculated curves. The interpolating curves usually coincide with calculations at a pressure beyond (10–20) TPa and a compression ratio of higher than $\sigma = 4.0$–4.5.

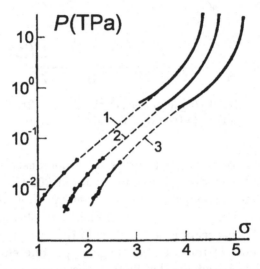

Fig. 6.6. Extrapolation of Hugoniots of rocks to the high-pressure range described by the Thomas–Fermi approximation: (1) dolomite ($\sigma - 0.4$); (2) slate; (3) granite ($\sigma + 0.4$).

7

Compression of
organic solids

The interest in organic materials is driven by several factors, one of which is their extensive application to devices operating under extreme conditions of high pressure and temperature, and information about their thermodynamic parameters in this region is essential. The second field of their application, which is no less important, is high-pressure chemistry. It is known that shock triggers various chemical transformations in many organic materials. Their nature is not clear, and investigation of physical processes in these reactions is an urgent problem addressed by researchers.

The primary topic is a classification of Hugoniots in accordance with classes of organic materials aiming to separate their common and distinctive features. Besides, it is virtually impossible to study all types of organic materials, therefore some classification of their properties is of great importance.

Generally speaking, few organic materials have been studied under shock compression in comparison to other solids, some of which were reviewed in previous sections. Therefore any new research on them is very interesting.

We measured the shock compressibility of six organic acids and three anhydrides [112] and also four polymers and one saturated hydrocarbon [46, 113–117]. They belong to the following groups: the succinic $HOOC(CH_2)_2COOH$ and glutaric $HOOC(CH_2)_3COOH$ acids are saturated dibasic acids; palmitic acid $C_{13}H_{27}COOH$ is monobasic; maleic acid $HOOCCH_2COOH$ is a nonsaturated diacid; tartaric or dihydroxysuccinic acid is $HOOC(CHOH)_2COOH$; and phthalic acid is orthophenylenecarboxylic diacid. The anhydrides were phthalic, succinic $(CH_2CO)_2O$, and myristic $(C_{13}H_{28}CO)_2O$. Tetracosane has the formula $C_{24}H_{50}$ $(C_nH_{2n+2}, \; n = 24)$, polyethylene $(-CH_2-CH_2-)$, polytetrafluorethylene $(-CF_2-CF_2-)_n$, polymethylmethacrylate $(C_5H_8O_2)_n$, and paraffin is a mixture of heavy saturated hydrocarbons, C_nH_{2n+2}, with $n > 20$.

Figure 7.1 demonstrates the similarity of Hugoniots of succinic,

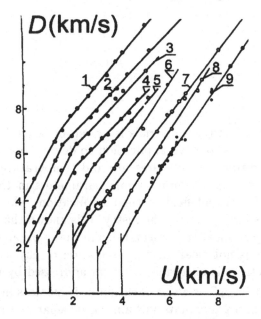

Fig. 7.1. Hugoniots of acids and some organic solids: (1) maleic acid $(D + 2)$; (2) glutaric acid $(D + 1)$; (3) succinic acid; (4) succinic anhydrate $(U + 0.5)$; (5) tartaric acid $(U + 1.0)$; (6) polyethylene $(U + 2)$; (7) Plexiglass $(U + 2)$; (8) Teflon $(U + 3)$; (9) paraffin $(U + 4)$ (full circles show measurements of tetracosane).

glutaric, and tartaric acids. The former two belong to one group, the difference between succinic and tartaric acids is that two hydrogen atoms in the former are replaced with hydroxyl groups. The Hugoniot of maleic acid has a slightly different slope for its second portion.

Measurements of the compression of succinic anhydride (its Hugoniot is close to that of the acid), of Teflon, polyethylene, Plexiglass, and paraffin are plotted in the same diagram. Since a lot of experimental data on $D-U$ Hugoniots of these four materials have been accumulated, we present only the data by reported by Russian groups, which by and large coincide with measurements by other researchers.

Most of the measurements of the Plexiglass Hugoniot, like those of other organic materials, were performed in laboratory conditions, but two experiments were conducted during underground nuclear tests. In the first experiment a pressure of 0.5 TPa was generated, which is beyond the range of laboratory facilities. The aim was, naturally, to extend the range of measured parameters (see Table 3.6). The second experiment, on the contrary, was performed under relatively low pressures ($P \simeq 6.0$ GPa) and was designed to test the reproducibility of the Plexiglass Hugoniot with respect to the sample thickness. As in the case of minerals and rocks, the question was about the dimensional effect, but in this experiment we suggested chemical reactions leading to the disintegration of the original structure into smaller groups, rather than a first-order transition. We assumed that the processes may not be completed in laboratory conditions, but in a full-scale experiment they should be completed *a fortiori*. In this case the Hugoniot should be shifted with respect to the laboratory curve.

Both experiments [11] used the standard impedance-matching technique. The shock generated by an underground explosion travelled across rock mass, an aluminum block with a thickness of about 100 mm, and a Plexiglass sample about 80 mm thick. Signals were collected from electric pins and recorded by memory oscilloscopes. The data are plotted in Figs. 5.8, 7.1.

It is noteworthy that the Plexiglass Hugoniot beyond the shock velocity $D \simeq 5.0$ km/s is a straight line in the $D - U$ coordinates with a slope $D'_U \simeq 1.25$. This means that most chemical transformations are completed at this velocity (the respective pressure is about 6.0 GPa). At lower parameters the Plexiglass Hugoniot is parabolic. Since the curves obtained in both laboratory and full-scale experiments coincide, the reactions of chemical dissociation are sufficiently fast and occur, most probably, on the shock front. Therefore the front is a generator of chemical reactions, as in the case of minerals it was a generator of denser phases in materials.

Note the shape of the Hugoniot of phthalic acid and its anhydride (Fig. 7.2), which are the only structures built around the benzene ring. In compression of benzene-based substances [115, 118] (see also Section 8.4) a cusp is detected on $D - U$ (and also on $P - \rho$) Hugoniots, and its shape indicates a first-order phase transition. A similar feature is observed on the phthalic acid Hugoniot. The curves of the acid and its anhydride are al-

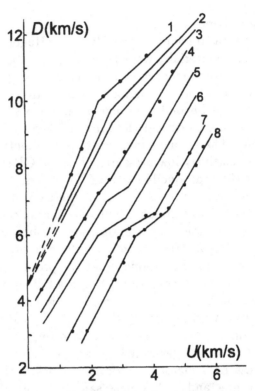

Fig. 7.2. Hugoniots of acids and anhydrates. Anhydrides: (1) myristic
$(D + 3)$; (2) butyric $(D + 3)$; (3) caproic $(D + 3)$; (8) phthalic $(D - 1,$
$U + 1.5)$. Acids: (4) palmitic $(D + 1.5)$; (5) butyric $(D + 1)$; (6) acetic
$(D + 0.5)$; (7) phthalic $(D - 1, U + 1)$.

most identical.

Now let us discuss the Hugoniots of the anhydrides of the series
$(C_nH_{2n+1}CO)_2O$. Of the three substances whose Hugoniots are
given in Fig. 7.2, only the anhydride of myristic acid is solid $(n =$
28), the other two (butyric with $n = 4$ and caproic with $n = 6$) are
liquids (see Section 8.5). The Hugoniots with higher n correspond
to higher shock velocities. All the three Hugoniots have almost
equal speeds of sound s_0 and the slope of the second portion drops
with n.

The Hugoniot of palmitic acid is compared to those of liquid
(butyric and acetic) acids (see also Section 8.5), which are lower
members of the series $C_nH_{2n+1}COOH$. The Hugoniots are much
alike, the only difference is that the particle velocity of the cusp

decreases with n, and it is almost invisible in the Hugoniot of palmitic acid, although the slopes of the second portions of $D - U$ curves are approximately equal for all materials.

To summarize this section, note that, by and large, Hugoniots of substances of one homologous series are similar, and the curves of acids and their anhydrides are also close to each other. The common feature of all compounds is the cusp on their Hugoniots, which indicates, most probably, that chemical reactions of decomposition of structures are triggered by the shock front.

8

Liquids

Since the number of liquids is practically unlimited, and they should be classified into numerous groups, which are often homologous series, include various solutions, mixtures, etc., it is impossible to study their properties individually. It is advisable to study representatives of various groups and extrapolate the measurements to other liquids of each group. A priori it was clear that many groups of liquids should behave similarly, and their properties would be described by similar Hugoniots. On the other hand, it was natural to admit that some liquids might have individual features and should be studied in dedicated experiments. A massive effort to study properties of liquids has been undertaken in the past ten to fifteen years, although on a lesser scale than the research on solids. Nonetheless, dozens of different liquids have been investigated, and some common features can be derived from these results.

Naturally, attention was primarily focused on water, which is the most abundant and important liquid on the Earth. The first measurements of its compression date back to 1957–58 [119, 120]. Measurements of its parameters were repeated with the aim of refining the results and extending the range of measured parameters of water under shock compression [8, 17, 28, 46, 47, 72].

The other groups of liquids subjected to shock compression were:

- saturated aqueous solutions of salts–halides, sulphates, and thiosulphates [121];
- organic liquids [28, 46, 47, 115, 122, 123];
- saturated hydrocarbons (alkanes) $C_n H_{2n+2}$ ($n = 6, 7, 10, 13, 14, 16$);

- nonsaturated hydrocarbons of the ethylene series (alkenes) $C_n H_{2n}$ (n=6–8) and of the acetylene series (alkynes) $C_n H_{2n-2}$ (n=6, 8);
- aromatic hydrocarbons, namely, benzene and its derivatives, such as toluene, nitrobenzene, aniline, and styrene;
- acids $C_n H_{2n+1} COOH$ (n=0, 1, 3, 5);
- anhydrides $(C_n H_{2n+1} CO)_2 O$ (n=3, 5);
- alcohols $C_n H_{2n+1} OH$ (n=2, 4–9);
- telomerized alcohols $H(CF_\alpha CF_\alpha)_n CH_\alpha OH$ (α=2, n=1–3);
- ketones, namely acetone and cyclohexanone.

Let us discuss measurements of these liquids.

8.1   Shock compression of water and aqueous solutions of salts

The basic aim of this research was to construct the equation of state needed for modelling propagation of shock in water, including sea water. The second problem of equal importance was to describe the compression of solutions using Hugoniots of water. Data on compression of water, including its solid state (ice), were also required to model the giant planets of the Solar System, some comets, etc.

The water Hugoniot shown in Fig. 8.1 consists of two rectilinear portions with different slopes joining at $D \simeq 5.0$ km/s. The slope of the first portion $D'_U = 1.7$, of the second portion 1.25. The cause of the cusp is a transformation of the pattern of bonds between water molecules under these conditions, although no consistent interpretation of this transition has been proposed.

The Hugoniot of ice ($\rho_{00} = 0.915$ g/cm^3) and two Hugoniots of 'porous' water (i.e., snow, $\rho_{00} = 0.6$ and 0.35 g/cm^3) are also given in the diagram. Since the maximum pressure at which ice melts is 0.2 GPa, the Hugoniots of porous states characterize the compressibility of the liquid phase. We cannot rule out generation of other modifications of water, including solid ones, at lower pressures.

The ice Hugoniot is practically identical to that of water. The Hugoniots of snow are rectilinear lines with close slopes ($D'_U \simeq 1.43$). This slope indicates that at high pressures the slopes of porous and solid Hugoniots should equalize. Judging by

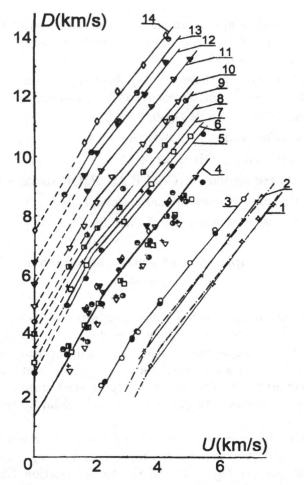

Fig. 8.1. Hugoniots of water, ice, and solutions of salts: (1) porous ice (snow, $\rho_{00} = 0.35$ g/cm^3, $D - 1$, $U + 0.5$); (2) porous ice ($\rho_{00} = 0.6$ G/cm^3, $D - 1$, $U + 1$); (3) ice ($\rho_{00} = 0.915$ g/cm^3, $D - 1$, $U + 1$); (4) water. Solutions: (5) NaCl ($D - 1$); (6) KCl ($D + 1.5$); (7) NaI ($D + 2$); (8) KI ($D + 2.5$); (9) CsI ($D + 3$); (10) CsBr ($D + 3.5$); (11) KBr ($D + 4$); (12) ZnSO$_4$ ($D + 5$); (13) ZnCl$_2$ ($D + 5.5$); (14) Na$_2$S$_2$O$_3$ ($D + 6$).

their shapes, this equality should be achieved at shock parameters higher than in these experiments ($D \geq 10$ km/s).

The dash-dotted lines show alternative shapes of snow Hugoniots.*

---

* In this version, all the four Hugoniots are approximately parallel lines with close slopes.

Further investigation of compressibility at a velocity $D \geq 2$ km/s is required to find out their exact positions.

Parameters of laboratory experiments are limited to $P = 100$ GPa, $D = 12$ km/s. Higher pressures were achieved in nuclear tests [8, 18] (see Table 5.1, Fig. 5.8). The agreement between laboratory and full-scale measurements is fairly good.

Figure 8.2 shows measurements of water compression plotted in $P - \sigma$ coordinates (left scale) and calculations of tempera-

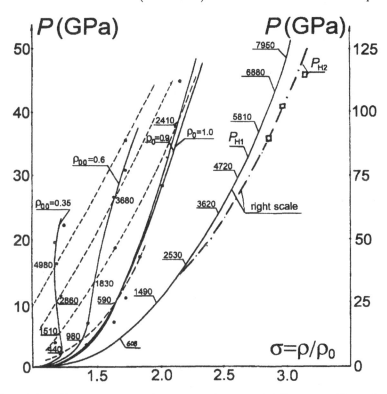

Fig. 8.2. $P - \sigma$ curves of water and ice: solid lines are Hugoniots; dashed lines are calculated expansion isentropes [72]; dash-dotted lines are double-compression Hugoniots. Numbers on the curves are calculated temperatures in kelvin.

ture (right scale) using the equation of state [72] (see Section 3.6, Eq. (3.7)). The parameters in the function $\eta(P)$, which was defined by the formula

$$\eta(P) = a[1 - \exp(-\alpha P)] + bP \exp(-\beta P), \quad a, \alpha, b, \beta = \text{const},$$

were determined from positions of experimental points for porous ice ($\rho_{00} = 0.35$ and 0.6 g/cm^3) on the $P - V$ plane and from the points on the double compression Hugoniot of water. The reference was the water Hugoniot with $\rho_0 = 1$ g/cm^3.

In Fig. 8.1 the water Hugoniot is compared to Hugoniots of saturated aqueous solutions of salts [121]. The Hugoniots of solutions lie in a relatively narrow band about the water Hugoniot and reproduce its shape fairly accurately. On the first portion $D'_U \simeq 1.8$, at $D \simeq 5$ km/s the slope changes to 1.2–1.3. It is remarkable that $D-U$ Hugoniots of aqueous solutions are separated from the water Hugoniot by ΔD, which is approximately equal to the difference between the speeds of sound in these liquids.

In other words, the Hugoniot of any solution can be plotted by taking the water Hugoniot, the speed of sound determined by conventional acoustic techniques, and plotting a line originating from the point corresponding to the speed of sound, s_0, parallel to the water Hugoniot. Hence we draw the following conclusions:

– firstly, Hugoniots of solutions with different salt concentration will be similar to the water Hugoniot and lie between the $D - U$ curve for water and the saturated solution;

– secondly, it seems reasonable to assume that Hugoniots of aqueous solutions of other salts (at least, of the same classes as those studied in our experiments) will also be similar to the water Hugoniot and shifted by the difference in the speed of sound, Δs_0;

– thirdly, irregularities on the $D - U$ curves of solutions are controlled by the properties of water, but not of solutes.

And the final remark. In previous sections we demonstrated that the additivity rule can be applied to Hugoniots of many materials. Solutions, which are homogeneous, two-component systems, cannot be classified with any group of materials in which the additivity rule is valid. So it is interesting to check whether the additivity rule is applicable to this case.

We found out that Hugoniots derived from Hugoniots of water and alkali-halide salts using the additivity rule conformed well to the experimental data. At the maximum pressure $P \sim 80$ GPa the difference between the calculated and measured density was within 3%. Therefore we conclude that the additivity rule yields a satisfactory description of all studied saturated solutions (and, naturally, of solutions with lower concentrations) and, probably, of solutions of other salts of these classes.

8.2 Saturated hydrocarbons (alkanes)

Six liquid alkanes–hexane C_6H_{14}, heptane C_7H_{16}, decane $C_{10}H_{22}$, tridecane $C_{13}H_{28}$, tetradecane $C_{14}H_{30}$, and hexadecane $C_{16}H_{34}$ have been studied [115]. Hexane was also studied in detail outside Russia [118]. The measurements are given in Fig. 8.3. The diagram also shows for comparison the data for one solid alkane–tetracosane $C_{24}H_{50}$. All the Hugoniots are linear and alike. The

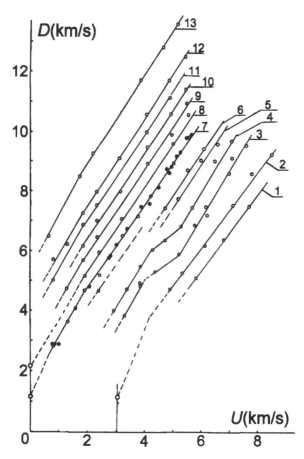

Fig. 8.3. Hugoniots: Alkynes: (1) hexyne ($U + 3.5$); (2) octyne ($U + 3$). Alkenes: (3) octene ($U + 2$); (4) heptene ($U + 1.5$); (5) hexene-2 ($U + 1$); (6) hexene ($U + 0.5$). Alkanes: (7) hexane; (8) heptane ($D + 1$); (9) decane; (10) tridecane ($D + 3$); (11) tetradecane; (12) hexadecane ($D + 5$); (13) tetracosane ($D + 3$).

curves of materials with a higher density, ρ_0, lie at higher shock velocities, D. All the curves belong to a narrow band in the diagram and can be approximated by similar formulas.

An analysis of coefficients a_1 and a_2 in the equation $D = a_1 + a_2 U$ for alkanes yielded their relations to the initial density, ρ_0:

$$a_1(\rho_0) = 5.59 - 13.6\rho_0 + 11.76\rho_0^2, \qquad a_2(\rho_0) = 1.34 + 0.16\rho_0.$$

These equations allow us to plot Hugoniots of any alkanes whose initial density is within the studied density range of 0.66–0.92 g/cm^3.

8.3   Nonsaturated hydrocarbons–alkynes and alkenes

Four substances of the ethylene series (alkenes)–two hexenes C_6H_{12}, heptene C_7H_{14}, and octene C_8H_{16}–and two hydrocarbons of the acetylene series (alkynes)–hexyne C_6H_{10} and octyne C_8H_{14}–were studied [122] (Fig. 8.3).

The Hugoniots of the best studied alkenes–octene and heptene–have jump-like discontinuities of their particle velocity and density. Let us recall that alkenes, unlike saturated organic compounds (alcohols, alkanes, anhydrides) with linear or more complex molecules, have double bonds between carbon atoms. It is possible that the discontinuities on their $D - U$ curves are due to breaking of these double bonds.

In the next section we shall see that $D - U$ curves of aromatic hydrocarbons, which also include double bonds, have similar discontinuities. Discontinuities on $D - U$ curves usually indicate a first-order phase transition to a denser phase with a jump in density. In this case jumps are most probably due to chemical transformations in materials in which double bonds in molecules are broken.

The $D - U$ Hugoniots of alkenes are similar: they all have steep first portions and less steep second portions. It is highly probable that the coefficients a_1 and a_2 defining the shapes of these curves can be related to the initial densities, as in the case of alkanes. But the amount of available data is not sufficient to derive these relations.

8.4   Aromatic hydrocarbons

A typical representative of these hydrocarbons is benzene C_6H_6. Its compressibility was measured by several research groups [46, 47, 115, 124]. The benzene $D - U$ Hugoniot has a jump (Fig. 8.4). The natural question was how hydrocarbons with a

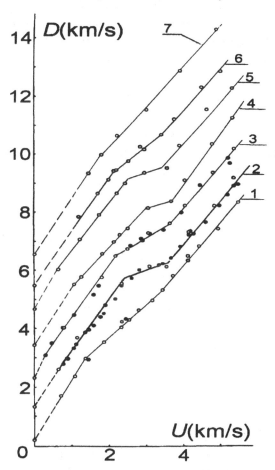

Fig. 8.4. Hugoniots of aromatic hydrocarbons and ketones. Ketones: (1) acetone $(D-1)$; (7) cyclohexanone $(D+5)$. Hydrocarbons: (2) benzene; (3) toluene $(D + 1)$; (4) styrene $(D + 3)$; (5) aniline $(D - 1)$; (6) nitrobenzene $(D + 2)$. Full circles show data by foreign researchers.

benzene ring and one hydrogen atom replaced with an atomic group would behave under shock. Therefore toluene $C_6H_5CH_3$ [115, 118], nitrobenzene $C_6H_5NO_2$, aniline $C_6H_5NH_2$, and styrene

$C_6H_5CH=CH_2$ [47, 124] were studied (toluene was also investigated by Yakusheva et al. [125]).

The results are given in Fig. 8.4. All the Hugoniots are very close (they are shifted along the ordinate axis in the diagram), they have almost equal slopes (especially on the first portions) and jumps. These jumps occur at close velocities and pressures of the transition $P \simeq$ 13–15 GPa. This feature is, apparently, caused by chemical reactions in which benzene rings are broken. Note that such transformations were not detected in hydrocarbons, such as alkanes, with linear molecules and ordinary bonds.

8.5   Acids and anhydrides

The studied acids have linear molecules with ordinary bonds [122]. Like aromatic hydrocarbons and alkenes, they have jumps on their $D-U$ curves. The jump amplitudes vary between 0.9 km/s (acetic acid) and $\simeq$ 0.5 km/s (formic acid). The Hugoniots have approximately equal slopes on both first (below the jump) and second portions.

Note that the speed of sound in most of the studied liquids [115, 122] (Fig. 8.5) was measured by detecting the travel time of a weak acoustic pulse between two electrodes with a selected separation between them.

The number of experimental points for anhydrides is not sufficient to plot definitely their $D - U$ curves, therefore two versions are given in the diagram. In one version of the anhydride Hugoniots they are close to the curves of respective acids.

8.6   Alcohols

Ten different alcohols, including three telomerized alcohols, were studied [46, 118, 122] (Fig. 8.6). All the Hugoniots are almost identical and have the shape of a broken line with a steep first portion ($D'_U \simeq 2.0$) and a less steep second portion ($D'_U \simeq 1.5$). The slope is changed at $D \simeq 4$ km/s. It is natural to assume the similarity of properties of alcohols from the proximity of their initial densities and sound velocities. To be exact, there is a correlation between the positions of $D - U$ curves and their initial densities, as in the case of alkanes. But the differences between their positions are so small that it is impossible to describe this

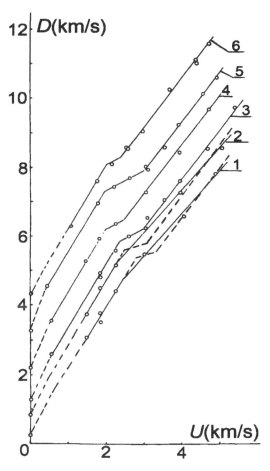

Fig. 8.5. Hugoniots of acids and anhydrides: (1) butyric anhydride $(D - 1)$; (2) caproic anhydride $(D - 0.5)$; (3) caproic acid; (4) butyric acid $(D + 1)$; (5) acetic acid $(D + 2)$; (6) formic acid (D_3).

correlation within the group of alcohols qualitatively. The averaged Hugoniot of the alcohols is described by the equation

$$D = 2.05 + 1.46U, \quad \bar{\rho}_0 = 0.835 \text{ g/cm}^3 \ (1.8 \text{ km/s} < U < 6 \text{ km/s}).$$

8.7 Ketones

The Hugoniots of two studied ketones – cyclohexanone and acetone [46, 118, 122] – are given in Fig. 8.4.

In studying aromatic hydrocarbons and alkenes, which have

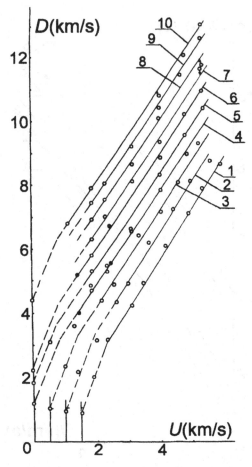

Fig. 8.6. Hugoniots of alcohols: (1) trihydrododecafluorheptanol
$(U + 1.5)$; (2) trihydrooctafluorpentanol $(U + 1)$; (3) trihydrotetrafluor-
propanol $(U+0.5)$; (4) ethyl alcohol; (5) butyl alcohol $(D+0.5)$; (6) amyl
alcohol $(D + 1)$; (7) hexyl alcohol $(D + 1.5)$; (8) heptyl alcohol $(D + 2)$;
(9) octyl alcohol $(D + 2.5)$; (10) nonyl alcohol $(D + 3)$.

double carbon bonds, we noted that double bonds in organic materials indicate the presence of jumps on their Hugoniots. This is
clearly seen on the Hugoniots of cyclohexadiene C_6H_8, cyclohex-
ene C_6H_{10}, and cyclohexane C_6H_{12}, whose densities and particle
velocity jumps drop with the decreasing number of double bonds
and are replaced with a change in the slope between the first and
second portions of the Hugoniots. This observation is confirmed
by the Hugoniot of cyclohexanone (Fig. 8.4), which has only single

bonds between carbon atoms and whose Hugoniot is a broken line of two rectilinear portions.

The second representative of ketones is acetone. Its $D - U$ Hugoniot consists of three rectilinear portions, including a 'spread' jump of the particle velocity.

9
Conclusion

To end this review, let us summarize the results of decades of research on the shock compression of condensed matter carried out in Russia.

Metals and numerous metallic compounds have been the subjects of the most intense research. The only group of these materials not studied under shock pressure is phosphides, which are relatively rare.

The number of experiments with minerals and oxides has been fairly large. Since the number of classes of these materials is enormous, the experimental data have been generalized to predict Hugoniots of similar materials in a wide range of shock parameters.

Research on shock compression of liquid condensed materials was started much later, but has been intense over recent years. However, the most interesting discoveries in this field related, in particular, to fast chemical reactions are expected in the late 1990s.

In conclusion let us consider some urgent tasks in the field of the shock compression of materials with a view to applying the results to the derivation of equations of state.

Primarily, this is further research into the compression of various materials at maximum pressures available in laboratory experiments. The aim is to obtain Hugoniots of solid (i.e., with a density $\rho_0 = \rho_{0,cr}$), porous ($\rho_0 < \rho_{0,cr}$), and superporous ($\rho_0 \ll \rho_{0,cr}$) materials.

Another important task is to develop a technique for measuring the compressibility of heated and precompressed vapors of materials, first of all, metals, with an initial density $\rho_0 < 0.1$ g/cm^3. The behavior of melted materials with initial temperatures $T >$ $1500 - 1700^0$ C should be also studied. Expansion Hugoniots with solid and highly porous initial states are needed to investigate states in the liquid–gas transition region and to verify wide-range equations of state.

The region of thermodynamic parameters of compressed materials between the curve of cold compression and the Hugoniot remains *terra incognita*. Research in this field may yield a lot of interesting results.

One urgent topic is the development of techniques for measuring the temperature not only of transparent, but of any materials under shock compression. These techniques will not only yield a new parameter directly involved in equations of state, but also allow the researchers to detect transitions to various states of aggregation under shock compression.

Techniques for measuring strength and viscosity of shock compressed materials are far from perfect. They are needed to determine the lower boundary of the applicability of the hydrodynamic model, on which the interpretation of shock measurements is based. The model presumes that the compression is triaxial since the applied load is fast and the material has no time to expand in directions perpendicular to the shock propagation axis. It seems reasonable to assume that this is the case under a pressure of tens of gigapascals. But what will happen at a lower pressure and where is the reasonable boundary for this assumption? Accurate measurements of strength and viscosity parameters of compressed solids should also yield information required to resolve this problem.

X-ray diffraction techniques, which have been developed in recent years for measurements under shock compression, are also very promising. It is obvious that measurement of lattice parameters of a shock compressed material is not an easy job. More interesting is the application of this technique to dynamic experiments. Phase transitions and their kinetics, structural transformations, melting, studies of $P - T$ diagrams, direct measurements of the density of materials under shock (using lattice parameters)– this list of problems demanding X-ray structural measurements is far from complete.

Naturally, there are a lot of minor problems that have not been ultimately solved in fifty years of dynamic experiments, and the researchers working in this field are well aware of them. Nonetheless, I would like to mention two of them, which are quite similar. These are the phase transitions in the graphite–diamond and quartz–stishovite systems. The experiments with them performed both in Russia and other countries have produced a lot of interesting data, but sometimes the results are controversial and cannot be easily interpreted. The problem cannot be solved without reproducing old and performing new experiments.

Owing to the comprehensive ban on nuclear explosions, the possibility of measurements under pressure higher than 3–5 TPa has been lost, although the problems which could be and should be solved using these experiments are numerous.

One of them is to detect oscillations in the shock amplitude at very high pressures when the compression energy sequentially exceeds ionization levels of the tested material. This topic was addressed in some experiments discussed in this review. The ultimate solution, however, can be found only when laboratory techniques generating pressures higher than 5 TPa and measuring material parameters with a high precision are available. Hope is now vested in powerful laser facilities which are being designed and whose radiation should generate shocks with the required amplitude in test materials.

Several theoretical models based on fundamental principles have been developed to describe states of highly compressed materials, and each of them yields calculations conforming to experimental data with a higher or lower degree of accuracy. Their common flaw, as it were, is that their accuracy declines considerably at lower pressures. To the best of our knowledge, some researchers are trying to improve these models and to extend their range to lower pressures. May they succeed in this feat.

References

[1] Al'tshuler, L.V., Krupnikov, K.K., Moiseev, B.N., Zhuchikhin, V.N., and Brazhnik, M.I. (1958). Dynamic compressibility and equation of state of iron under high pressures, *Zh. Eksp. Teor. Fiz.* **32** (4), 874.

[2] Al'tshuler, L.V., Krupnikov, K.K., and Brazhnik, M.I. (1958). Dynamic compressibility of metals under a pressure of up to four hundred thousand to four million atmospheres, *Zh. Eksp. Teor. Fiz.* **34**, 886.

[3] (a) Zhernokletov, M.V., Simakov, G.V., Sutulov, Yu.N., and Trunin, R.F. (1995). Expansion isentropes of Al, Fe, Mo, Pb, and Ta, *Teplofiz. Vys. Temp.* **33** (1), 40. (b) Bugaeva, V.A., Evstigneev, A.A., and Trunin, R.F. (1996). Expansion Hugoniots of Cu, Fe, and Al. Analysis of calculations, *Teplofiz. Vys. Temp.* **34** (5), 684.

[4] Al'tshuler, L.V., Moiseev, B.N., Popov, L.V., Simakov, G.V., and Trunin, R.F. (1968). Relative compressibility of iron and lead under a pressure of 31–34 Mbar, *Zh. Eksp. Teor. Fiz.* **54** (3), 785.

[5] Trunin, R.F., Simakov, G.V., Podurets, M.A., Moiseev, B.N., and Popov, L.V. (1971). Dynamic compressibility of quartz and quartzite under high pressure, *Izv. Akad. Nauk SSSR, Fiz. Zemli.* No. 1, 13.

[6] Trunin, R.F., Podurets, M.A., Moiseev, B.N., Simakov, G.V., and Popov, L.V. (1969). Relative compressibility of copper, cadmium, and lead under high pressure, *Zh. Eksp. Teor. Fiz.* **36** (4), 1172.

[7] Trunin, R.F., Podurets, M.A., Simakov, G.V., Popov, L.V., and Moiseev, B.N. (1972). Experimental test of the Thomas–Fermi model for metals under high pressure, *Zh. Eksp. Teor. Fiz.* **62** (3), 1043.

[8] Podurets, M.A., Simakov, G.V., Trunin, R.F., Popov, L.V., and Moiseev, B.N. (1972). Compression of water by intense shock waves, *Zh. Eksp. Teor. Fiz.* **62** (2), 170.

[9]   Trunin, R.F., Podurets, M.A., Simakov, G.V., Popov, L.V., and
      Sevast'yanov, A.G. (1995). New data on the compressibility of
      aluminum, Plexiglass, and quartz obtained under intense shock
      waves generated by nuclear explosions, *Zh. Eksp. Teor. Fiz.*
      108 (3), 851.

[10] Trunin, R.F., Simakov, G.V., Moiseev, B.N., Popov, L.V., and
      Podurets, M.A. (1969). About the existence of carbon metal
      phase under dynamic compression, *Zh. Eksp. Teor. Fiz.* **56** (4),
      1169.

[11] Trunin, R.F. (1994). Shock compressibility of condensed
      materials under intense shock generated by underground
      nuclear explosion, *Uspekhi Fiz. Nauk.* **164** (11), 215.

[12] Trunin, R.F., Simakov, G.V., Dudoladov, I.P., Telegin, G.S.,
      and Trusov, I.P. (1988). Compressibility of rocks under shock
      waves, *Izv. Akad. Nauk SSSR, Fiz. Zemli.* No. 1, 52.

[13] Trunin, R.F., Medvedev, A.B., Funtikov, A.I., Podurets, A.M.,
      Simakov, G.V., and Sevast'yanov, A.G. (1989). Shock
      compressibility of porous iron, copper, and tungsten, and their
      equations of state in the terapascal range, *Zh. Eksp. Teor. Fiz.*
      95 (2), 632.

[14] Trunin, R.F., Podurets, M.A., Popov, L.V., Zubarev, V.N.,
      Bakanova, A.A., Ktitorov, V.M., Sevast'yanov, A.G., Simakov,
      G.V., and Dudoladov, I.P. (1992). Measurements of iron
      compressibility at 5.5 TPa, *Zh. Eksp. Teor. Fiz.* **102** (3), 1433.

[15] Trunin, R.F., Podurets, M.A., Popov, L.V., Moiseev, B.N.,
      Simakov, G.V., and Sevast'yanov, A.G. (1993). Shock
      compressibility of iron under a pressure of up to 10 TPa
      (100 Mbar), *Zh. Eksp. Teor. Fiz.* **106** (6), 2189.

[16] Trunin, R.F., Il'kaeva, L.A., Podurets, M.A., Popov, L.V.,
      Pechenkin, B.V., Prokhorov, L.V., Sevast'yanov, A.G., and
      Khrustalev, V.V. (1994). Shock compressibility of iron, copper,
      lead, and titanium under a pressure of up to 20 TPa, *Teplofiz.*
      Vys. Temp. **32** (5), 692.

[17] Avrorin, E.N., Vodolaga, B.K., Volkov, L.P., Vladimirov, A.S.,
      and Chernovolyuk, B.T. (1980). Shock compressibility of lead,
      quartzite, aluminum, and water at 100 Mbar, *Pis'ma v Zh.*
      Eksp. Teor. Fiz. **31** (12), 727.

[18] Avrorin, E.N., Vodolaga, B.K., Voloshin, N.P., Kovalenko, G.V.,
      Simonenko, V.A., and Chernovolyuk, B.T. (1987). Experimental
      observation of shell restructuring effects on Hugoniots of
      condensed materials, *Zh. Eksp. Teor. Fiz.* **93** (2), 613.

[19] Volkov, L.P., Voloshin, N.P., Vladimirov, A.S., Nogin, V.N., and
      Simonenko, V.A. (1980). Shock compressibility of aluminum at
      10 Mbar, *Pis'ma v Zh. Eksp. Teor. Fiz.* **31** (11), 623.

[20] Simonenko, V.A., Voloshin, N.P., Vladimirov, A.S., Nagibin, A.P., Popov, V.A., Sal'nikov, V.A., and Shoidin, Yu.A. (1985). Absolute measurements of aluminum shock compressibility at 1 TPa, *Zh. Eksp. Teor. Fiz.* **88** (4), 1452.

[21] Podurets, M.A., Ktitorov, V.M., Trunin, R.F., Popov, L.V., Matveev, A.Ya., Pechenkin, B.V., and Sevast'yanov, A.G. (1994). Shock-wave compression of aluminum to 1.7 TPa, *Teplofiz. Vys. Temp.* **32** (6), 952.

[22] Ragan III, C.E., Silbert, M.G., and Diven, B.C. (1977). Shock compression of molybdenum to 2.0 TPa by means of nuclear explosion, *J. Appl. Phys.* **48** (7), 2860.

[23] Ragan III, C.E. (1980). Ultrahigh-pressure shock-wave experiments, *Phys. Rev. A* **21** (2), 458.

[24] Ragan III, C.E. (1984). Shock-wave experiments at threefold compression, *Phys. Rev. A* **29** (3), 1391.

[25] Ragan III, C.E. (1982). Shock compression measurements at 1 to 7 TPa, *Phys. Rev. A* **25** (6), 3360.

[26] Mitchell, A.C., Nellis, W.I., and Holmes, N.C. (1984). Shock impedance match experiments in aluminum and molybdenum between 0.1–2.5 TPa (1–25) Mbar, in *Shock Wave in Condensed Matter-1983*, ed. I.R. Asay, R.A. Graham, G.K. Straub (Elsevier Science Publishers, B. V.), Chapter II.

[27] Fuqian, Jing. (1990). *Shock Wave Physics Research in China*, ed. S.C. Schmidt, I.N. Johnson, and L.W. Davidson (Elsevier Science Publishers, B. V.).

[28] Trunin, R.F. (1992). *Properties of Condensed Materials under High Pressure and Temperature*, [in Russian], ed. R.F. Trunin (Izd. VNIIEF, Arzamas-16).

[29] Al'tshuler, L.V., Trunin, R.F., Krupnikov, K.K., and Panov, N.V. (1996). Laboratory explosive facilities for studying compression of materials due to shock waves, *Uspekhi Fiz. Nauk* **166** (5), 575.

[30] Trunin, R.F., Panov, N.V., and Medvedev, A.B. (1995). Shock compressibility of iron, aluminum, and tantalum under terapascal pressure, *Khim. Fiz.* **14** (2-3), 97.

[31] Trunin, R.F., Panov, N.V., and Medvedev, A.B. (1995). Compressibility of iron, aluminum, and tantalum under a shock pressure of 1–2.5 TPa, *Pis'ma v Zh. Eksp. Teor. Fiz.* **62** (7), 572.

[32] Al'tshuler, L.V., Bakanova, A.A., Dudoladov, I.P., Dynin, E.A., Trunin, R.F., and Chekin, B.S. (1981). Hugoniots of metals. New data, statistical analysis, and general trends, *Zh. Prikl. Mekh. Tekh. Fiz.* No. 2, 3.

[33] Al'tshuler L.V. and Bakanova, A.A. (1968). Electronic structure
      and compressibility of metals under high pressure, *Uspekhi Fiz.
      Nauk* **96** (2), 193.

[34] Kalitkin, N.N. and Kuz'mina, L.V. (1975). *Tables of
      Thermodynamic Functions of Materials at High Energy Density,*
      [in Russian], (Preprint of the Institute of Problems of
      Mechanics, Akad. Nauk SSSR) No. 3.

[35] Kopyshev, V.P. (1977). Thermodynamics of nuclei of
      monoatomic materials, *Chislennye Metody Mekhaniki Sploshnoi
      Sredy* **8** (6), 54.

[36] Nikiforov, A.F., Novikov, B.G., and Uvarov, V.B. (1989).
      Modified Hartree–Fock–Slater model and its application to
      derivation of equations of state at high temperature, in
      *Mathematical Modelling. Physico-Chemical Properties of
      Materials*, [in Russian], (Nauka, Moscow).

[37] Sin'ko, G.V. (1983). Application of self-consistent field method
      to thermodynamic functions of electrons in elementary
      materials, *Teplofiz. Vys. Temp.* **2**, No. 6, 1041.

[38] Nikiforov, A.F., Novikov, V.G., and Uvarov, V.B. (1979).
      Modified Hartree–Fock–Slater model for matter at given
      temperature and density, *Voprosy Atomnoi Nauki i Tekhniki:
      Metodiki i Programmy Chislennykh Uravnenii Zadach
      Material'noi Fiziki*, No. 4(6), 16.

[39] Nikiforov, A.F., Novikov, V.G., Trukhanov, S.K., and Uvarov,
      V.B. (1990). Derivation of equation of state of aluminum based
      on modified Hartree–Fock–Slater model, *Voprosy Atomnoi
      Nauki i Tekhniki: Matematicheskoe Modelirovanie Fizicheskikh
      Protsessov*, No. 3, 62.

[40] Trunin, R.F., Podurets, M.A., Simakov, G.V., Popov, L.V., and
      Sevast'yanov, A.G. (1994). Shock compressibility of
      molybdenum at 1.4 TPa, *Teplofiz. Vys. Temp.* **32** (5), 786.

[41] Kalitkin, N.N.. (1989). Models of matter under extreme
      conditions, in *Mathematical Modelling: Physico-Chemical
      Properties of Materials*, [in Russian], (Nauka, Moscow).

[42] Simonenko, V.A. (1990). Experimental measurements of
      equation of state for metals at 200 Mbar and comparison with
      Thomas–Fermi and other theoretical models, *High-Pressure
      Research* **5**, 826.

[43] Mitchell, A.C., Nellis, W.J., Moriarty, J.A., Heinle, R.A.,
      Holmes, H.C., Tipton, R.E., and Repp, G.W. (1991). Equation
      of state of Al, Cu, Mo, and Pb at shock pressure of up to
      2.4 TPa (24 Mbar), *J. Appl. Phys.* **69** (5), 2981.

[44] Skidmore, I.C. and Morris, E. (1962). Experimental equation of
      state data for uranium and its interpretation in the critical

region, in *Thermodynamics of Nuclear Materials*, [Proceedings of Symposium], (Vienna).

[45] Glushak, B.L., Zharkov, A.P., Zhernokletov, M.V., Ternovoi, V.Ya., Filimonov, A.S., and Fortov, V.E. (1989). Experimental study of thermodynamics of dense metal plasma at a high energy density, *Zh. Eksp. Teor. Fiz.* 96 (4), 1301.

[46] LASL. (1980). Shock Hugoniot Data, in *Los Alamos Series on Dynamic Material Properties*, ed. S.P. Marsh (Univ. California Press, Berkeley–Los Angeles–London).

[47] Compendium of Shock-Wave Data. (1977). ed. M.Van Thiel (UCRL-55108).

[48] Trunin, R.F., Medvedev, A.B., Funtikov, A.I., Podurets, M.A., Simakov, G.V., and Sevast'yanov, A.G. (1989). Shock compression of porous Fe, Cu, and W, and their equations of state under terapascal pressures, *Zh. Eksp. Teor. Fiz.* 95 (2), 631.

[49] Model', I.Sh., Narozhnyi, A.T., Kharchenko, A.I., Kholin, S.A., and Khrustalev, V.V. (1985). Equations of state of graphite, aluminum, titanium, and iron at a pressure higher than 13 Mbar, *Pis'ma v Zh. Eksp. Teor. Fiz.* 46 (6), 270.

[50] Vladimirov, A.S., Voloshin, N.P., Nogin, V.A., Petrovtsev, A.V., and Simonenko, V.A. (1984). Shock compressibility of aluminum under a pressure $P \geq 1$ Gbar, *Pis'ma v Zh. Eksp. Teor. Fiz.* 39 (2), 69.

[51] Krupnikov, K.K., Brazhnik, M.I., and Krupnikova, V.P. (1962). Shock compression of porous tungsten, *Zh. Eksp. Teor. Fiz.* 42 (3), 675.

[52] Zel'dovich, Ya.B. (1957). Derivation of equation of state from mechanical measurements, *Zh. Eksp. Teor. Fiz.* 32 (2), 1577.

[53] Kormer, S.B., Funtikov, A.I., Urlin, V.D., and Kolesnikova, A.N. (1962). Dynamic compression of porous materials and equation of state with a variable specific heat at high temperatures, *Zh. Eksp. Teor. Fiz.* 42 (3), 626.

[54] Trunin, R.F., Simakov, G.V., Sutulov, Yu.N., Medvedev, A.B., Rogozkin, B.D., and Fedorov, Yu.E. (1989). Compression of porous materials under shock waves, *Zh. Eksp. Teor. Fiz.* 96 (3), 1024.

[55] Simakov, G.V. and Trunin, R.F. (1991). Shock compression of superporous silicon dioxide, *Izv. Akad. Nauk SSSR, Fiz. Zemli*, No. 11, 72.

[56] Trunin, R.F. and Simakov, G.V. (1993). Shock compression of nickel of very low density, *Zh. Eksp. Teor. Fiz.* 103 (6), 2180.

[57] Minstev, V.B., and Fortov, V.E. (1979). Electric conductance of zenon under supercritical conditions, *Pis'ma v Zh. Eksp. Teor. Fiz.* 30 (7), 401.

[58]   Zaporozhets, Yu.B., Mintsev, V.B., Fortov, V.E., and Batovskii, O.M. (1984). Reflection of laser light from shock compressed xenon plasma, *Pis'ma v Zh. Eksp. Teor. Fiz.* **40** (10), 1339.

[59]   Al'tshuler, L.V., Brusnikin, S.E., and Marchenko, A.I. (1989). Determination of Grüneisen coefficient of highly nonideal plasma, *Teplofiz. Vys. Temp.* **27**, 636.

[60]   Al'tshuler, L.V., Kormer, S.B., Bakanova, A.A., and Trunin, R.F. (1960). Equations of state of aluminum, copper, and lead in the high-pressure range, *Zh. Eksp. Teor. Fiz.* **38** (3), 790.

[61]   Al'tshuler, L.V., Bakanova, A.A., and Trunin, R.F. (1962). Hugoniots and zero isothermal curves of seven metals at high pressures, *Zh. Eksp. Teor. Fiz.* **42** (1), 91.

[62]   Lennard-Jones, J.E. and Devonshire, A.F. (1937). *Proc. Roy. Soc. A* **163**, 3.

[63]   Latter, R. (1966). Temperature behaviour of the Thomas–Fermi statistical model for atoms, *Phys. Rev.* **99** (6), 1854.

[64]   Urlin, V.D. (1965). Melting under superhigh shock presure, *Zh. Eksp. Teor. Fiz.* **49** (2), 485.

[65]   Sapozhnikov, A.T. and Pershina, A.V. (1979). Semiempirical equation of state for metals in wide ranges of density and pressure, *Voprosy Atomnoi Nauki i Tekhniki*, No. 4(6), 47.

[66]   Glushak, B.L., Gudarenko, L.F., Styazhkin, Yu.M., and Zherebtsov, V.A. (1991). Semiempirical equation of state for metals with variable electron specific heat, *Voprosy Atomnoi Nauki i Tekhniki, Ser. Matematicheskoe Modelirovanie Fizicheskikh Protsessov*, No. 1, 32.

[67]   Al'tshuler, L.V., Bushman, A.V., Zhernokletov, M.V., Zubarev, V.N., Leont'ev, A.A., and Fortov, V.E. (1980). Relief isentropes and equations of state of metals at high energy density, *Zh. Eksp. Teor. Fiz.* **78** (2), 741.

[68]   Kormer, S.B., Urlin, V.D., and Popova, L.T. (1961). Interpolation equation of state and its application to measurements of shock compression of metals, *Fiz. Tverd. Tela* **3** (7), 2131.

[69]   Bushman, A.V. and Fortov, V.E. (1983). Modelling equations of state of materials, *Uspekhi Fiz. Nauk* **140** (2), 177.

[70]   Glushak, B.L., Gudarenko, L.F., and Styazhkin, Yu.M. (1991). Semiempirical equation of state for metals with variable thermal capacity of nuclei and electrons, *Voprosy Atomnoi Nauki i Tekhniki, Ser. Matematicheskoe Modelirovanie Fizicheskikh Protsessov*, No. 2, 57.

[71]   Zubarev, V.N., Podurets, M.A., Simakov, G.V., Popov, L.V., and Trunin, R.F. (1974). Shock compression and equation of state of copper in the high-pressure range, in *Proceedings of 1st*

All-Union Symposium on Pulsed Pressure, [in Russian], (VNIIFTRI, Moscow).

[72] Bakanova, A.A., Zubarev, V.N., Sutulov, Yu.N., and Trunin, R.F. (1975). Thermodynamic properties of water under high pressure and temperature, *Zh. Eksp. Teor. Fiz.* **66** (3), 1099.

[73] Medvedev, A.B. (1990). Modelling equation of state taking into account evaporation, in *Topics in Nuclear Science and Technology. Ser. Theoretical and Applied Physics*, [in Russian], No. 1, 23.

[74] Kopyshev, V.P. (1971). Thermodynamic model of dense liquid, *Zh. Prikl. Mekh. Tekh. Fiz.* **1**, 119.

[75] Medvedev, A.B. (1992). Modelling equation of state taking into account evaporation, ionization, and melting, in *VANT, Ser. Theoretical and Applied Physics*, [in Russian], No. 1, 12.

[76] Kopyshev, V.P. and Khrustalev, V.V. (1980). Equation of state of hydrogen at a pressure of up to 10 Mbar, *Zh. Prikl. Mekh. Tekh. Fiz.* **1**, 122.

[77] Trunin, R.F., Belyakova, M.Yu., Zhernokletov, M.V., and Sutulov, Yu.N. (1991). Shock compression of metallic alloys, *Izv. Akad. Nauk SSSR, Fiz. Zemli*, No. 2, 99.

[78] Bakanova, A.A., Bugaeva, V.A., Dudoladov, I.P., and Trunin, R.F. (1995). Shock compression of nitrides and carbides of metals, *Izv. Akad. Nauk SSSR, Fiz. Zemli*, No. 6, 58.

[79] Pavlovskii, M.N. (1971). Shock compression of diamond, *Fiz. Tverd. Tela* **13** (3), 893.

[80] Adadurov, G.A., Dremin, A.N., Pershin, S.V., Rodionov, V.N., and Ryabinin, Yu.N. (1962). Shock compression of quartz, *Zh. Prikl. Mekh. Tekh. Fiz.*, No. 4, 81.

[81] Wackerle, I. (1962). Shock-wave compression of quartz, *J. Appl. Phys.* **33** (3), 922.

[82] Al'tshuler, L.V., Trunin, R.F., and Simakov, G.V. (1965). Shock compression of periclase and quartz, and composition of the lower Earth mantle, *Izv. Akad. Nauk SSSR, Fiz. Zemli*, No. 10, 1.

[83] Trunin, R.F., Simakov, G.V., and Podurets, M.A. (1971). Compression of porous quartz by intense shock waves, *Izv. Akad. Nauk SSSR, Fiz. Zemli*, No. 2, 33.

[84] Podurets, M.A., Simakov, G.V., Telegin, G.S., and Trunin, R.F. (1981). Polymorphism of silicon dioxide under shock and equation of state of coesite and stishovite, *Izv. Akad. Nauk SSSR, Fiz. Zemli*, No. 1, 1.

[85] Podurets, M.A., Simakov, G.V., and Trunin, R.F. (1976). Equilibrium among phases in shock compressed quartz and kinetic of phase transition, *Izv. Akad. Nauk SSSR, Fiz. Zemli*, No. 7, 3.

[86]	Podurets, M.A., Popov, L.V., Sevast'yanov, A.G., Simakov, G.V., and Trunin, R.F. (1976). Effect of sample dimension on position of silicon dioxide Hugoniot, *Izv. Akad. Nauk SSSR, Ser. Fiz. Zemli*, No. 11, 56.

[87]	Podurets, M.A., Simakov, G.V., and Trunin, R.F. (1990). Transformation of stishovite to a denser phase on the second branch of quartz Hugoniot, *Izv. Akad. Nauk SSSR, Fiz. Zemli*, No. 4, 30.

[88]	Holmes, N.C. (1993). Shock compression of low-density foams, in *High-Pressure Science and Technology-1993*, ed. S.C. Schmidt, J.W. Shaner, G.A. Samara, M. Ross, part 1, 153.

[89]	Vil'danov, V.G., Gorshkov, M.M., Slobodenyukov, V.M., and Rashkovan, E.Kh. (1995). Shock compression of porous quartz under pressures of up to 100 GPa, in *J. Press, Proceedings of the 1995 APC Topical Conference on Shock Compression of Condensed Matter*, (Seattle, Wash.).

[90]	Furnish, M.D., and Ito, E. (1995). Experimental measurements of shock properties of stishovite, in *J. Press, Proceedings of the 1995 Topical Conference on Shock Compression of Condensed Matter*, (Seattle, Wash.).

[91]	Al'tshuler, L.V. and Petrunin, A.P. (1961). X-ray diffraction measurements of light materials under obliquely colliding shock waves, *Zh. Eksp. Teor. Fiz.* **31** (6), 717.

[92]	Simakov, G.V., Pavlovskii, M.N., Kalashnikov, N.G., and Trunin, R.F. (1974). Shock compression of twelve minerals, *Izv. Akad. Nauk SSSR, Fiz. Zemli*, No. 8, 11.

[93]	Simakov, G.V. and Trunin, R.F. (1980). Shock compression of minerals, *Izv. Akad. Nauk SSSR, Fiz. Zemli*, No. 2, 77.

[94]	Trunin, R.F., Simakov, G.V., and Podurets, M.A. (1974). Shock compression of porous rutile, *Izv. Akad. Nauk SSSR, Fiz. Zemli*, No. 12, 13.

[95]	Bugaeva, V.A., Podurets, M.A., Simakov, G.V., Telegin, G.S., and Trunin, R.F. (1979). Dynamic compressibility and equation of state of minerals with rutile structure, *Izv. Akad. Nauk SSSR, Fiz. Zemli*, No. 1, 28.

[96]	Kalashnikov, N.G., Pavlovskii, M.N., Simakov, G.V., and Trunin, R.F. (1973). Dynamic compressibility of minerals of calcite group, *Izv. Akad. Nauk SSSR, Fiz. Zemli*, No. 2, 23.

[97]	Trunin, R.F., Gon'shakova, V.I., Simakov, G.V., and Galdin, N.E. (1965). Investigation of rocks under high shock pressure and temperature, *Izv. Akad. Nauk SSSR, Fiz. Zemli*, No. 9, 1.

[98]	Simakov, G.V., Podurets, M.A., and Trunin, R.F. (1973). New data about compressibility of oxides and fluorides, and hypothesis of uniform Earth composition, *Dokl. Akad. Nauk SSSR* **211** (6), 1330.

[99] Al'tshuler, L.V., Pavlovskii, M.N., Kuleshova, L.V., and Simakov, G.V. (1963). Investigation of alkali metal halides under high shock pressure and temperature, *Fiz. Tverd. Tela* 5 (1), 279.

[100] Kormer, S.B., Sinitsin, M.V., Funtikov, A.I., Urlin, V.D., and Blinov, A.V. (1964). Compressibility of five ionic compounds under a pressure of up to 5 Mbar, *Zh. Eksp. Teor. Fiz.* 47 (4), 1202.

[101] Lin-gun, Lin and Basset, W.A. (1973). Compression of Ag and phase transformation of NaCl, *J. Appl. Phys.* 44 (4), 1475.

[102] Piermarini, G.J. and Block, S. (1975). Ultrahigh pressure diamond-anvil cell and several semiconductor phase transitions pressures in relation to the point pressure scale, *Rev. Sci. Instrum.* 46 (8), 973.

[103] Al'tshuler, L.V., Kuleshova, L.V., and Pavlovskii, M.N. (1960). Dynamic compressibility, equation of state, and electric conductance of sodium chloride at high pressure, *Zh. Eksp. Teor. Fiz.* 39 (1), 16.

[104] Alder, B. (1966). Physical experiments with intense shock waves, in *Solids under high pressure*, ed. W. Paul and D. Warshauer (Mir, Moscow).

[105] Kuleshova, L.V. and Pavlovskii, M.N. (1976). Phase transition in NaCl under shock loading, *Fiz. Tverd. Tela* 18 (2), 573.

[106] Al'tsheller, L.V. (1965). Applications of shock waves to high pressure physics, *Uspekhi Fiz. Nauk* 85 (2), 179.

[107] Podurets, M.A., Trunin, R.F., Shcherbinskaya, N.P., and Moiseev, B.N. (1995). Phase transitions in rock salt under an intense shock wave generated by an underground nuclear explosion, *Zh. Eksp. Teor. Fiz.*, to be published.

[108] Trunin, R.F. (1986). Compressibility of various materials under high shock pressure, *Izv. Akad. Nauk. SSSR, Fiz. Zemli*, No. 2, 26.

[109] Hughes, D.S. and McQueen, R.C. (1957). Density of basic rocks at very high pressures, *Trans. Amer. Geophys. Union* 3, 21.

[110] Telegin, G.S., Antoshev, V.G., Bugaeva, V.A., Simakov, G.V., and Trunin, R.F. (1980). Calculation of Hugoniots of rocks, *Izv. Akad. Nauk SSSR, Fiz. Zemli*, No. 5, 22.

[111] Al'tshuler, L.V. and Sharipdzhanov, I.I. (1971). Additive equations of state of silicates at high pressures, *Izv. Akad. Nauk SSSR, Fiz. Zemli*, No. 3, 11.

[112] Trunin, R.F., Zhernokletov, M.V., Dorokhin, V.V., and Sychevskaya, N.P. (1992). Compression of solid organic acids and anhydrides under shock waves, *Khim. Fiz.* 11 (4), 557.

[113] Kalashnikov, N.G., Kuleshova, L.V. and Pavlovskii, M.N. (1972). Shock compression of polytetrafluorethylene at a

pressure of up to 1.7 Mbar, *Zh. Prikl. Mekh. Tekh. Fiz.*, No. 4, 187.

[114] Bakanova, A.A., Dudoladov, I.P., and Trunin, R.F. (1965). Compression of alkali metals under intense shock waves, *Fiz. Tverd. Tela* **7** (6), 1615.

[115] Trunin, R.F., Zhernokletov, M.V., Kuznetsov, N.F., and Sutulov, Yu.N. (1989). Dynamic compressibility of saturated and aromatic hydrocarbons, *Khim. Fiz.* **8** (4), 539.

[116] Alekseev, Yu.F., Al'tshuler, L.V., and Krupnikova, V.P. (1971). Shock compression of two-component paraffin-tungsten mixtures, *Zh. Prikl. Mekh. Tekh. Fiz.*, No. 4, 152.

[117] Dremin, A.N. and Karpukhin, I.A. (1960). Method for determination of Hugoniots of dispersed materials, *Zh. Prikl. Mekh. Tekh. Fiz.*, No. 3, 91.

[118] Dick, R.D. (1970). Shock wave compression of benzene, carbon disulphide, carbon tetrachloride, and liquid nitrogen, *J. Chem. Phys.* **52** (12), 6021; (1979). Shock compression data for liquids. I. Six hydrocarbon compounds, *J. Chem. Phys.* **71** (8), 3203.

[119] Walsh, I.M. and Rice, M.H. (1957). Dynamic compression of liquids from measurements of strong shock waves, *J. Chem. Phys.* **26** (4), 815.

[120] Al'tshuler, L.V., Bakanova, A.A., and Trunin, R.F. (1958). Phase transformations in water compressed by intense shock waves, *Dokl. Akad. Nauk SSSR* **121** (1), 67.

[121] Trunin, R.F., Zhernokletov, M.V., Kuznetsov, N.F., Simakov, G.V., and Shutov, V.V. (1987). Dynamic compressibility of aqueous solutions of some salts, *Izv. Akad. Nauk SSSR, Fiz. Zemli*, No. 12, 37.

[122] Trunin, R.F., Zhernokletov, M.V., Kuznetsov, N.F., Radchenko, O.A., Sychevskaya, N.P., and Shutov, V.V. (1992). Compression of liquid organic materials under shock waves, *Khim. Fiz.* **11** (3), 424.

[123] Nellis, W.I., Ree, F.H., Van Thiel, M., and Mitchell, A.C. (1981). Shock compression of liquid carbon monoxide and methane to 90 GPa (900 Kbar), *J. Chem. Phys.* **75** (6), 3055.

[124] Yakushev, V.V., Dremin, A.N., Nabatov, S.S., and Shunin, V.M. (1979). Physical properties and transformation of benzene under dynamic pressure of up to 30 GPa, *Fiz. Goreniya i Vzryva*, No. 2, 132.

[125] Yakusheva, O.B., Yakushev, V.V., and Dremin, A.N. (1977). Correlation between loss of transparency and anomalies on curves of shock compression in carbonic compounds under high dynamic pressure, *Zh. Fiz. Khimii* **61** (7), 1657.

Index

Additivity rule, 95, 102, 134

Compressibility
  absolute, 9, 12
  relative, 8, 10
Compression, 26, 32, 56, 136

Detectors
  light, 58
  photoelectric, 51
  piezoceramic, 65
  pin, 17, 34, 57, 65

Earth core, 103, 105, 112
Equations
  additive, 135
  caloric, 77
  gas, 80, 83, 90
  Hugoniot, 78
  multiple regression, 134
  of state (EOS), 76, 79, 85, 87
    'co-volume', 89
    interpolation, 79, 83
    Mie–Grüneisen, 77, 89
    ROSA, 86, 87
    Van-der-Waals, 89

Full-scale experiments, 10

Grüneisen coefficient, 75, 78, 83
  electron analogue, 68, 79, 85

Models

ACTEX, 42
  Hartree–Fock–Slater (HFS), 41
  INFERNO, 42
  modified HFS (MHFS), 40, 42
  self-consistent field (SCF), 40
  TFCK, 40, 80, 83
  Thomas–Fermi (TF), 40, 83

Nuclear explosion, 7, 22, 33, 53

Parameters
  kinematic shock, 3, 21
  thermodynamic, 1, 3, 5
    elastic ('cold'), 71, 77
    shock compression, 71
    thermal, 72, 78, 80, 82, 83, 86

Shock generators
  plane, 16
  spherical configuration, 18

Techniques
  γ-reference, 11, 49
  deceleration, 5, 34
  Doppler, 12
  impedance-matching, 6, 22, 91
  inverse impedance-matching, 24

Velocity
  impactor, 5, 35
  particle, 3, 25
  shock, 3, 25
  sound, 26, 70, 140